D1441888

Life at the Limits
Organisms in extreme environments

We are fascinated by the seemingly impossible places in which organisms can live. There are frogs that can freeze solid, worms that dry out and bacteria that survive temperatures over 100 °C. These organisms have an extreme biology, which involves many aspects of their physiology, ecology and evolution. In this captivating account, the reader is taken on a tour of extreme environments, and shown the remarkable abilities of organisms to survive a range of extreme conditions, such as high and low temperatures and desiccation. This book considers how organisms survive major stresses, and what extreme organisms can tell us about the origin of life and the possibilities of extraterrestrial life.

DAVID WHARTON is a Senior Lecturer in the Department of Zoology at the University of Otago, New Zealand, where his research has centred on the extraordinary survival abilities of animals. His interest in extreme environments has been stimulated by visits to the Antarctic, and by a year spent at the British Antarctic Survey in Cambridge, as a Royal Society Guest Research Fellow. In 1997, his contribution to science was recognised by the award of Doctor of Science, by the University of Bristol.

He is also an expert in the use of light and electron microscopy in natural history filmmaking, and was a principal instigator of the Diploma in Natural History Filmmaking and Communication at the University of Otago, in partnership with Natural History New Zealand.

Life at the Limits

Organisms in extreme environments

DAVID A. WHARTON
Department of Zoology
University of Otago
New Zealand

CAMBRIDGE
UNIVERSITY PRESS

PUBLISHED BY THE PRESS SYNDICATE OF THE UNIVERSITY OF CAMBRIDGE
The Pitt Building, Trumpington Street, Cambridge, United Kingdom

CAMBRIDGE UNIVERSITY PRESS
The Edinburgh Building, Cambridge CB2 2RU, UK
40 West 20th Street, New York, NY 10011–4211, USA
477 Williamstown Road, Port Melbourne, VIC 3207, Australia
Ruiz de Alarcón 13, 28014 Madrid, Spain
Dock House, The Waterfront, Cape Town 8001, South Africa

http://www.cambridge.org

First published 2002

Printed in the United Kingdom at the University Press, Cambridge

Typeface Trump Mediaeval 9.5/15 pt.   *System* QuarkXPress™ [SE]

A catalogue record for this book is available from the British Library

Library of Congress Cataloguing in Publication data

ISBN 0 521 78212 0 hardback

This book is dedicated to my children Rebecca and Simon. May the Earth they inherit not be made more extreme by the actions of humanity.

Contents

Preface

I hope you will excuse me starting by talking about myself but it might help to explain some of my background and how I came to write this book. When I was doing my bachelor's degree in Zoology at the University of Bristol in England, one of my final year projects was on the structure of the eggshell, and the hatching mechanisms of the eggs, of a nematode parasite of cockroaches that rejoices in the splendid name of *Hammerschmidtiella diesingi* (named after a German nematologist, Hammerschmidt). Nematodes are a group of worm-like invertebrate animals. The eggshell turned out to have complex systems of pores and spaces in its outer layers. I stayed on at Bristol to do a PhD that followed up this work and looked at the eggshells of a variety of parasitic nematodes. The nematode eggshell is one of the most resistant of all biological structures. In some cases, it can even survive immersion in concentrated sulphuric acid. The eggshell has a layer of lipid, which restricts the exchange of materials between the egg and its environment. The eggs lose water very slowly when exposed to desiccation, enabling the larvae enclosed within them to survive the total loss of water from the egg. The pores in the eggshell of *H. diesingi* are involved in this process.

The free-living stages of some parasitic nematodes (those developing after the hatching of the egg), and of some free-living nematodes, show extraordinary abilities to survive extreme environmental stresses. They will tolerate freezing, complete desiccation and exposure to chemicals that are fatal to most other organisms (such as the fixatives used to kill and stabilise specimens for microscopy). Most of my career has been involved in trying to determine the mechanisms by which nematodes survive these extreme insults. I have gradually transformed from being a parasitologist (who studies the biology of parasites) into an environmental physiologist (who studies the interactions between organisms and their environment). When I moved to New Zealand in 1985, I had the opportunity to visit the Antarctic. The organisms that live in the rare terrestrial Antarctic sites that will support life are living in one of the most extreme environments on Earth. Experiencing such conditions myself stimulated my interest in extreme life even more.

Nematodes, of course, are not the only organisms that are able to survive in extreme environments. Since about the 1960s, we have become aware that a variety of microorganisms can flourish in all sorts of extreme conditions and that life is possible where it was previously thought to be impossible. This has broadened our understanding of the nature of life on Earth and the possibility of life elsewhere in the universe. In this book, I attempt to cover all types of organisms (animals, plants and microbes), the different types of extreme environments and the different aspects of their biology in these environments. I will also try to develop a framework for deciding what is extreme for life and for thinking about the biology of extreme organisms. This has been a large task – I will leave it to you to decide how well I have succeeded! Most people who have written about organisms in extreme environments are microbiologists or biochemists. As a zoologist, I am perhaps aware of a wider (or at least a different) range of organisms and approaches. I thus hope to bring a different perspective to the subject.

The book is written to be understandable by a non-expert (and hopefully even those with little background in biology or science). I have tried to explain things as I've gone along, but there is also a glossary to clarify unfamiliar terms. In order to keep the readability of the text, I have not formally cited references to original research papers. Instead, I have tried to include in the bibliography those that will enable the source of original research to be found and also mention in the text some of the major workers in the relevant areas. If any of my scientific colleagues feel that their work should have been directly cited, and was not, my apologies to them.

I would like to thank especially my wife Ann who, apart from supporting me during the writing of the book, read every chapter and, since she is not a scientist, helped me to make it more readable and understandable. Thanks also to my (now ex-) research student, Brent Sinclair, who also read every chapter and who was not shy about making suggestions for improvements. I would also like to thank those who read and commented on individual chapters and/or the book proposal: Carolyn Burns, Colin Townsend, Dev Nyogi, Bill Block, Don Gaff, Hans Ramløv, Craig Cary, Rick Lee and Graham Young. It was John Montgomery, of the University of Auckland, who started me thinking about resistance and capacity adaptations in relation to extreme environments. Jo Ogier has produced some wonderful illustrations and Ken Miller and Matthew Downes assisted me with those I prepared myself. Don Gaff and Craig Cary kindly supplied me with copies of their illustrations. The sources of material for illustrations, and relevant permissions, are acknowledged in the figure legends, where appropriate. Finally, I would like to thank my editors at Cambridge University Press, Alice Houston and Simon Mitton (and their team), for guiding me through the process of producing this book.

David A. Wharton
Dunedin
New Zealand

1   Introduction: extreme life

In 1989, I was lucky enough to visit the Antarctic for the first time, as part of the New Zealand Antarctic Research Programme. I was looking for nematodes, a group of worm-like invertebrate animals that live associated with the algae and moss that grow in the meltwater from snow and glaciers, and around the edges of lakes and small ponds. I visited various sites around the McMurdo Sound area of Antarctica, including the Dry Valleys which form the largest area of ice-free land on the continent. Parts of the Dry Valleys are called 'oases', because they support some visible signs of life. If you were expecting palm trees, however, you would be disappointed. Small patches of moss are as good as it gets in this part of Antarctica. The organisms that live here face some of the most extreme conditions experienced on Earth.

We all have ideas as to what might be normal and extreme environmental conditions. We might like our normal environment to be lying on the back lawn in the dappled sunlight with a gin and tonic. The heat of the desert (without the gin and tonic!) or the cold and wind of the Antarctic or Arctic might, in contrast, seem somewhat extreme. For other organisms (and even other humans), however, these places are home. In this book, I aim to explore the adaptations that have enabled organisms (including plants, animals and microbes) to live in situations that we might consider extreme and to try to develop a framework for thinking about organisms and extreme environments.

WHAT IS EXTREME?

Many organisms experience environmental conditions that seem to us to be relatively 'normal', but some are able to survive or even thrive in conditions which we might regard as 'extreme'. This judgment is based on our own experience of our environment. The great majority of organisms live permanently in the sea. We would find more than a brief immersion beneath the waters of the sea an extreme stress, unless we had special equipment, but what we would find extreme is normal for marine organisms. There is clearly a problem with defining extreme conditions by our own experience. Can we develop less subjective criteria for determining what might be normal and what might be extreme for an organism?

Measuring the responses of organisms to changes in environmental conditions might provide the tools we need. Let us use the effect of temperature as an example. The responses of organisms to temperature are complex. The simplest response to temperature is that there is an optimum, at which activity, growth rates and metabolism are greatest, and a range of temperatures that an organism will survive (Figure 1.1). As the temperature increases or decreases from the optimum, the metabolism of the organism decreases. If the temperature becomes more extreme, the organism may display heat or cold stupor, in which movement becomes disorganised and normal processes are disrupted. Close to the limit of the tolerable range of temperatures, the organism will display heat or cold coma and cease activity altogether. Once the temperature limits are exceeded, it will die. Establishing the temperatures at which these changes in activity occur for an organism may allow us to determine what is 'normal' and 'extreme' with respect to temperature for that organism. Defining these transition temperatures may, however, not be easy because the organism may respond to changing temperature by initiating biological responses that extend the survivable limits. In the case of low temperatures, the temperature at which metabolism ceases may not represent the point at which the organism dies.

There are important differences between the lethal effects of high and low temperatures. The damage caused by high temperature is destructive as proteins become denatured and other irreversible changes occur. The effect of low temperature may be rather different. As the temperature falls, metabolism slows (and if the temperature is low enough, it ceases) as the kinetic energy imparted to chemical reac-

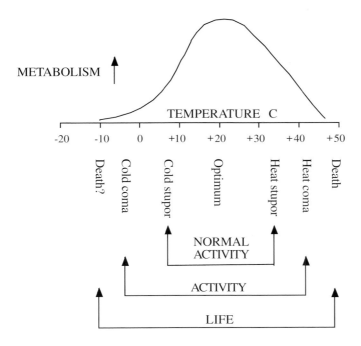

FIGURE I.I The responses to temperature in a hypothetical organism. At low temperatures, metabolism is undetectable. As the temperature increases, the rate of metabolism increases due to the increased kinetic energy supplied to reactions. Beyond the optimum temperature, however, metabolism slows and eventually ceases due to the damaging and lethal effects of high temperature. Changes in activity are associated with these changes in temperature. As the temperature increases or decreases from the optimum, the organism may become disorientated and normal processes disrupted (heat or cold stupor) and then cease altogether (heat or cold coma). Death may then result. These transitions define the ranges over which normal activity and life can occur.

tions decreases. This effect is potentially reversible. Death may result, however, from events such as irreversible changes in membrane function, although freezing, or the risk of freezing, is likely to be the major hazard. Freezing involves a change in the state of water within the organism from a liquid to a solid. This can be a sudden and violent event, and initiates a number of changes that may result in death, unless the organism has mechanisms which enable it to survive the stresses involved. The lethal effects of heat are unlikely to be due to a change of state in body water since, for most organisms, their upper lethal temperature is many degrees lower than the temperature at which water boils.

THE LIFE BOX

Just as there are ranges for survival and activity with respect to temperature, the same is true for other environmental variables (such as salinity, conditions of acidity or alkalinity, oxygen concentration etc.). We could measure conditions (temperature, pH, salinity etc.) in the environment adjacent to an organism. If we did this lots of times over the lifespan of the organism, we could determine the range of conditions it has experienced. Some organisms move around, experiencing changes in environmental conditions, and conditions themselves change with time. Plotting these measurements in multidimensional space would thus define the overall range of conditions that the organism has experienced (its 'life box'; Figure 1.2). I have experienced −26 °C in the Antarctic and +45 °C in the Australian desert and so my life box would extend between these two values for the parameter of temperature. If we did the measurements for all organisms of the same species, we could define the life box for that species, and thus the sort of habitats in which that species could live. If we did the measurements for all organisms of all species, we could define the life box for life in general. If conditions change such that an organism finds itself outside the life box for its species, it will die. If conditions are such that they are outside the life box for life in general, there will be no life. Ecologists call an organism's life box its 'ecological niche'. This is determined by both the physical characteristics of its environment (temperature, pH, water availability etc.) and its interactions with other organisms (predation, competition, disease, the food source it exploits etc.). I have used the term 'life box' to focus attention on the range of physical conditions that organisms can tolerate and because it might help us to identify what is extreme for organisms.

To decide what might be an extreme organism, it might be useful to think in terms of the life box for the majority of organisms or for the

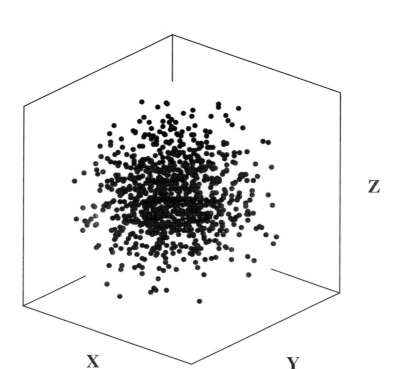

FIGURE 1.2 The life box of a hypothetical organism with respect to three environmental variables (x, y, z; e.g. temperature, pH, oxygen availability). Each point represents the conditions experienced and survived by the organism at a particular place and moment in time. The conditions experienced and survived at all places and times defines the life box for the organism or species of organism. If the conditions go beyond its life box, the organism dies.

majority of species of organism. An extreme organism would, therefore, be one that tolerates conditions beyond those tolerated by most organisms. The range of conditions for such an organism will be different from that for the majority of organisms, and it will have a life box which occupies a different theoretical space from that of most organisms (Figure 1.3). Such organisms have been called extremophiles (they love conditions which are far removed from the ordinary or average). The best-known examples are thermophilic bacteria that live associated with hot springs and deep-sea hydrothermal vents, where the temperatures they experience are much higher than those experienced

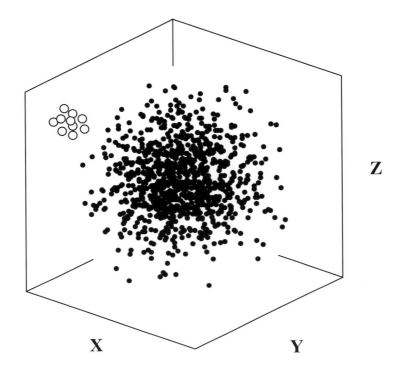

FIGURE 1.3 The life box of an extremophile (○) compared with that of the majority of organisms. For the extremophile, the range of conditions that it will survive is different from that survived by the majority of organisms. Its life box thus occupies a different theoretical space.

by most organisms. There are extremophiles that colonise other types of extreme environments – such as very saline habitats (halophiles), acidic or alkaline conditions (acidophiles, alkaliphiles), low temperatures (psychrophiles) and high pressures (piezophiles).

There is, however, another group of organisms that can be considered to be extreme. In terms of the conditions in which they can maintain activity, their life box is the same as, or considerably overlaps, the life box of the majority of organisms. However, when conditions change to the point where the organism can no longer sustain metabolic activity, rather than dying, they cease metabolism and enter into an ametabolic dormant state. When conditions return to normal, they

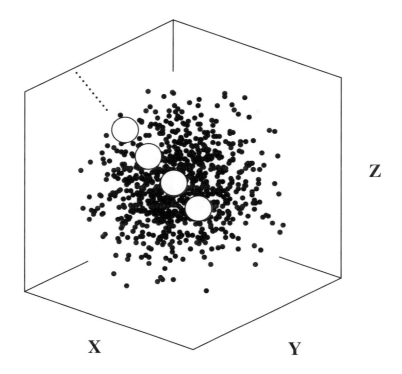

FIGURE 1.4 The life box of a cryptobiote, only here the symbols represent conditions under which the organism can metabolise rather than survive. Its life box is thus the same as, or considerably overlaps, the life box for the majority of organisms (see Figures 1.2 and 1.3). When conditions change to beyond those in which the cryptobiotic organism can metabolise (···O···), metabolism ceases, but the cryptobiote can resume metabolism once conditions become favourable again.

resume activity. In other words, if we think of life in terms of metabolism, they have the capacity to step outside their life box and to become active again once conditions become favourable (Figure 1.4). This phenomenon has been called cryptobiosis (hidden life), anabiosis (renewed life) or latent life. Latent life is perhaps the most appropriate term, since, in the latent state, the capacity for life is present but is not apparent. Cryptobiosis is, however, the most commonly used term. Some cryptobiotic organisms can enter this state at any stage in their life cycle. Others have special survival and dispersal stages, which act as lifeboats that carry organisms to new habitats, or which enable them to survive periods unsuitable for growth. This ability enables them to survive in space and time until conditions are favourable. These lifeboats include spores, eggs, cysts, seeds and resistant larval stages.

Cryptobiotic organisms can survive a variety of environmental stresses. Some can survive the complete loss of their body water. This phenomenon has been called 'anhydrobiosis' (life without water). Other types of cryptobiosis include cryobiosis (extreme cold), thermobiosis (heat), osmobiosis (osmotic stress, such as high salt concentrations) and anoxybiosis (lack of oxygen). A number of organisms enter a period of dormancy in which their activity levels are lowered in response to these types of environmental stresses. The lowering of metabolic rate may be to as little as 80 per cent of normal resting levels, but more typically to a point in the range of 5–40 per cent of resting levels. Cryptobiosis is distinguished from dormancy by resulting in a depression of metabolic rate to less than 1 per cent of resting levels or even ceasing altogether.

WHO WANTS TO LIVE FOREVER?

How long can organisms survive in a state of cryptobiosis? For nematodes, the record is 39 years. Rotifers (another group of microscopic invertebrate animals) have been revived from dried herbarium specimens that were 120 years old. Plant seeds can lie dormant for many years. The ancient English herbs weld and mullein grew from soil from a Carthusian priory that was closed during the dissolution of the monasteries by Henry VIII between 1536 and 1540. These 400-year-old seeds, the plants of which had not been seen in England since medieval times, grew after the soil containing them was brought to the surface after an archaeological dig. The oldest seed ever germinated is a 1300-year-old lotus seed from China.

Microorganisms hold the most remarkable records for longevity. Bacteria have grown from spores from a 118-year-old can of meat (canned veal taken on Parry's Arctic expedition, 1820–1830). Beer has been brewed from yeast isolated from a bottle of porter ale which was taken from the wreck of a sailing barge that lay off the coast of the English port of Littlehampton for 166 years. Somewhat more controversial are reports of bacteria that may be millions of years old. Bacteria have been isolated from rocks, salt deposits and permafrost (permanently frozen soil). Many have doubted these reports since it is difficult to prove that samples have not been contaminated with bacteria of more recent origin. The most convincing claims concern those from specimens that are naturally protected against contamination. Bacterial spores (from the genus *Bacillus*) were isolated from a bee preserved in amber estimated to be 20–40 million years old. The material was protected against contamination by the amber, the surface of which was carefully sterilised before the sample was taken. Bacteria (also a *Bacillus*) have been isolated from liquid inclusions enclosed within salt crystals and estimated to be 250 million years old. Like those isolated from amber, the material was protected from contamination by the salt crystal and extreme care was taken to prevent contamination during sampling. If the evidence for these sorts of claims holds up, there is no reason not to believe that bacterial spores can be immortal.

EXTREME LANGUAGE

I have mentioned a number of terms that describe organisms which grow in or survive extreme environmental conditions. Perhaps I'd better explain these terms a bit more before we go any further. There is an imposing terminology that has been developed by scientists working on different types of organisms and different types of environmental stress. Organisms that grow best under extreme conditions (the conditions for their optimal growth is much higher or lower than the average for most organisms) are referred to as being extremophilic. The ending '-philic' means 'loving' (from the Greek 'philia', meaning affection or fondness). Organisms that can survive extreme conditions but whose optimal growth conditions lie within the more normal range are referred to as being 'tolerant'. The extreme conditions result in a reduction in metabolism and a period of dormancy which the organism can survive. Where metabolism ceases altogether, organisms are called cryptobiotic. This means 'hidden life' and is derived from the Greek words for hidden ('kryptos', hide or conceal) and life ('biosis').

Terms which describe the responses to different environmental stress are derived by adding these endings to roots for the particular stress. The roots are: thermo- (heat, from the Greek 'therme'), cryo- or psychro- (cold, from 'kryos' for icy cold and 'psychros' for cold or frigid); anhydro- or xero- (desiccation, from 'anhydros' for waterless and 'xeros' for dry); piezo- or baro- (pressure, from 'piezo' for press and 'baros' for weight); halo- or osmo- (osmotic stress, from 'halos' for salt and 'osmos' for pushing); and acido- or alkali- (low and high pH). A lack of oxygen is referred to as anaerobic (without air) or anoxic (without oxygen). These terms are summarised in Table 1.1.

EXTENDING THE LIFE BOX

Extremophiles thrive in extreme environments, while cryptobiotes can survive extreme conditions until more moderate conditions return. There are, however, other responses to extreme conditions. Organisms may avoid the extreme conditions by migrating to more favourable ones. Snow geese avoid the cold of the Arctic winter by migrating south to more moderate conditions. Desert insects avoid desiccation and heat during the day by burrowing into the sand. Some organisms can modify their external or internal environment to make conditions more normal and less extreme. In the case of temperature, this type of response is mainly found in birds and mammals.

Most organisms are ectotherms – they are at the same temperature as that of their environment. Birds and mammals, however, are endotherms. They can generate their own heat and maintain a higher temperature within their bodies than that in their surrounding environment. This is achieved firstly by burning fuel (food) to generate heat through metabolism and secondly by mechanisms to reduce the loss of heat to the environment, such as through insulation (fur,

Table 1.1 *Some extreme terminology*

Stress	Required for growth *philic*	Tolerated and/ or slow growth *tolerant*	Metabolism ceases *biosis*
Cold *psychro* cryo	Psychrophilic	Psychrotolerant Cold tolerant	Cryobiosis
Heat *thermo*	Thermophilic	Thermotolerant	Thermobiosis
Pressure *baro* *piezo*	Piezophilic Barophilic	Piezotolerant Barotolerant	—
Osmotic *halo* *osmo*	Halophilic Osmophilic	Halotolerant Osmotolerant	Osmobiosis
Acid/alkali *acido* *alkali*	Acidophilic Alkaliphilic	Acidotolerant Alkalitolerant	—
No oxygen *anaero* *anoxy*	Obligate anaerobes	Facultative anaerobes	Anoxybiosis
Desiccation *anhydro* *xero*	Xerophilic	Xerotolerant	Anhydrobiosis

feathers and fat beneath the skin). These mechanisms enable birds and mammals to live in some very cold places.

Humans show the final response to extreme conditions. We modify our external environment to bring the conditions within the range we can survive. Other animals do this to a limited extent, by building nests and burrows, but our ability to modify our environment has enabled humans to colonise, or at least survive in, almost any place on Earth. If we get cold, we put on extra clothing or turn on a heater. We carry water with us into the desert and can even desalinate saltwater to provide freshwater for drinking or irrigation where it is otherwise in short supply. Where there is no oxygen, we carry it with us. Our ability to modify our environment has enabled us to survive at the tops of the highest mountains, in the depths of the sea, in the driest, hottest deserts, in the barren waste of the Antarctic polar plateau and even in space.

RESISTANCE AND CAPACITY ADAPTATIONS

Organisms may have two types of responses to adverse conditions. Resistance adaptations enable the organism to avoid or survive the stress until conditions become favourable again; capacity adaptations enable the organism to grow and reproduce under the harsh conditions. Although these terms were originally proposed for adaptations to temperature stress (in 1958 by Precht, a German physiologist), they could be applied to other stresses such as lack of oxygen, osmotic stress and desiccation. Precht considered capacity adaptations to be those that operated within the normal range of temperatures experienced by an organism, while resistance adaptations operated at extreme temperatures. I have extended Precht's meaning to make the terms apply to the response of organisms to extreme environments. For an organism displaying capacity adaptation to an extreme habitat, the extreme conditions *become* the normal range of conditions. The enzymes, membranes and other systems of the organism are optimised to operate at extremes (such as at very high or very low temperatures). If the conditions for the optimum growth of an organism which lives in a harsh environment are different from those of organisms that live in more benign environments, this would provide evidence for capacity adaptation: for example, if the optimum growth temperature of an Antarctic organism was lower than those of its relatives from warmer regions.

For an organism displaying resistance adaptation to extreme conditions, its normal range of conditions is the same as those of most other organisms. When faced with extreme conditions, outside the range where it can normally operate, it can survive until conditions become 'normal' again, whereas other organisms may die. Resistance adaptations often involve some sort of dormancy, such as hibernation or the production of a resting stage like a cyst, spore or seed. In response to conditions becoming extreme, or in response to changes which indicate that extreme conditions are on their way (such as the onset of winter), the organism becomes inactive. This involves a lowering of the metabolic rate of the organism, reaching its ultimate expression in cryptobiosis, in which there is no measurable metabolism. The lowering of metabolic rate may in itself provide some resistance to the adverse conditions, by reducing the rate at which food reserves are consumed. There may also be more specific mechanisms which provide protection against the stress – for example, the synthesis of cryoprotectants in response to low temperatures.

Extremophiles thus show capacity adaptation to extreme environments, while cryptobiotes show resistance adaptation. The difference between the two has been illustrated in Figures 1.2 and 1.3. The two types of adaptation are not mutually exclusive and many organisms show both capacity and resistance adaptations to extreme conditions. The extreme thermophilic bacterium *Pyrococcus furiosus*, an inhabitant of deep-sea hydrothermal vents, has an optimum growth temperature of 100 °C and the temperatures in which it can grow range from 70 °C to 105 °C. At temperatures below 70 °C, it becomes dormant. This bacterium thus shows capacity adaptation to high temperatures, but resistance adaptation when temperatures become too low for it.

SOME BASICS

For most of the rest of this chapter, I will introduce some physical and biological concepts that are important for understanding the ideas covered later in the book. Card-carrying biologists might wish to skim through this.

Normal and extreme conditions

What are normal and extreme conditions for life? As we have seen, temperature has important influences on organisms and both high and low temperatures may be considered extreme. Many terrestrial organisms maintain some sort of normal activity over the rather narrow range of 10 °C to 48 °C (a range of 38 °C). The lowest natural temperature recorded on Earth is −89.2 °C, at Vostok, Antarctica (temperatures within a fraction of a degree of absolute zero, −273 °C, have been achieved in laboratories). The highest natural temperatures at the Earth's surface occur associated with volcanoes. In the absence of geothermal activity, the highest reported shade temperature is 58 °C, recorded at Al'Azízíyah, Libya. This gives a range of recorded temperatures (in the absence of geothermal activity) of about 147 °C. The largest natural recorded temperature range at one location is 105 °C (at Verkhoyansk, Siberia). High temperatures (above 48 °C), low temperatures (below 0 °C) and large temperature ranges, particularly if they occur within a short period of time, may be considered extreme.

What are normal temperatures? The Earth is, on average, a cold place. More than two-thirds of the Earth's surface is covered by ocean and the temperature of most of the ocean stays close to 2 °C. Including the ocean depths, the polar ice caps and the land, four-fifths of the planet is below 5 °C all the time. What we might think of as 'normal' temperatures, say 10 °C to 30 °C, are really not normal at all but occur only in restricted parts of the world. Abundant life is associated with the warmer parts of the Earth (but not too warm!).

Organisms on land experience the weight of the air pressing down on them. We are so used to this that we are barely conscious of it unless it suddenly changes, as, for example, in a plane where our ears may pop as ascent or descent results in different pressures on either side of our eardrums. At sea level, the pressure is 1 kilogram per square centimetre or 1 atmosphere. Changes in this pressure may be experienced as a stress by an organism. Most organisms do not experience low pressure naturally. Even high mountains will not produce a lowering in pressure sufficient to stress an organism (although there may be a problem with a lack of oxygen). Some bacteria, nevertheless, can happily grow at low pressures, and even in a vacuum, which is an issue for the spoilage of vacuum-packed food.

Terrestrial organisms also rarely experience high pressures. Even the deepest mine does not produce a significant increase in pressure (at a depth of 1500 metres, the pressure is only about one-sixth higher than at the surface). In the ocean, however, changes in pressure with depth are significant. The hydrostatic pressure (the weight of water pressing on the organism) increases by 1 atmosphere for every 10 metres' depth. Even at the bottom of shallow coastal waters or a lake, the pressure is several times greater than that experienced at the surface. In the deepest parts of the ocean, some 11 kilometres deep, the pressure is 1100 atmospheres (that is, 1100 times that found at the surface). This represents a considerable stress for any organism that might live there (1100 kilograms of pressure per square centimetre – imagine five elephants standing on the tip of your finger!).

When some substances dissolve in water, they result in the splitting of water molecules (H_2O) to generate hydrogen (H^+) or hydroxyl (OH^-) ions. A high concentration of H^+ ions (and a low concentration of OH^- ions) produces an acid solution, whereas a low concentration of H^+ ions (and a high concentration of OH^- ions) produces an alkaline solution. The degree of acidity or alkalinity is indicated by the pH of the solution. This is a measure of the concentration of H^+ ions in the solution (on a logarithmic scale). Water is pH 7.0, which is considered neutral (neither acid nor alkaline), solutions with pH values lower than 7.0 are acidic and those with pH values higher then 7.0 are alkaline.

What is a normal pH? A neutral pH of 7.0 is usually thought of as normal, but, in fact, many natural sources of water (seas, lakes, rivers, soil water) are slightly acidic with a pH of 5.6. This is because carbon dioxide, from the atmosphere, dissolves in the water to produce a weak acid (carbonic acid). Despite living in slightly acidic conditions, the pH inside most cells is 7.7. Living cells control their internal pH at this level because it is optimum for the functioning of their enzymes – the biological catalysts which control the chemical reactions of organisms and enable them to grow, reproduce and maintain their organisation. Strong acids (like sulphuric acid) and alkalis (such as caustic soda) not only prevent the efficient functioning of enzymes but can also destroy the proteins, membranes and other structures which make up the body of the organism.

When substances dissolve in water, they impose other stresses on organisms, in addition to the effect that some of them have on the pH. When common salt (sodium chloride) is added to water, it dissolves; if more is added, more will dissolve, until no more can dissolve and the solution is saturated. As more salt is added, the concentration of salt in the solution increases, but, conversely, the concentration of water in the solution decreases (the amount of water in the container remains the same but the amount of salt is increasing and so the concentration of water is decreasing). When an organism is placed in a salty solution, if the concentration of water outside the organism is lower than that inside, water will leave the organism (unless it is able to prevent it from doing so) to try and restore the balance (a process called diffusion in which molecules, such as water, move from a region of high concentration to one of low concentration). The alternative is that the salt would diffuse into the organism's cells. However, the membranes of cells allow some substances to pass through them but not others (they are permeable to some substances but impermeable to others, and are referred to as being semi-permeable). The cell membrane is permeable to water but impermeable to salts (or at least more permeable to water than it is to salts). The diffusion of water across a semi-permeable membrane is called osmosis and a condition in the external environment that results in such movement is referred to as an osmotic stress.

When the salt and water concentrations inside and outside cells are equal, there will be no osmotic stress and we might perhaps think of this as being the normal (or, at least, unstressed) condition. This is true of many marine organisms whose internal fluids have the same osmotic concentration as the surrounding seawater. The cells of the organism will be under an osmotic stress if the internal and external conditions are not equal. There are two types of osmotic stress. A hyperosmotic stress is when the concentration of water outside the cell is lower than that inside (due to a higher salt concentration) and water leaves the cell. This dehydrates the cell and causes it to shrink. A hyposmotic stress is when the concentration of water outside the cell is higher than that inside (due to a lower salt concentration outside the cell) and water enters the cell. The entry of water due to a hyposmotic stress causes the cell to swell and, if the cell doesn't remove the excess water, it is in danger of bursting. Most organisms regulate the internal salt concentration (or osmotic concentration) of their cells so that it is slightly higher than that of the external environment. This creates a slight positive pressure within the cell (like a gently inflated balloon) which keeps the cell firm (turgid).

Organisms that live in seawater (in which the salt concentration is approximately equivalent to a 0.85 per cent solution of sodium chloride) will experience an hyposmotic stress if they are transferred into freshwater, and most will die. The pressure generated by the osmotic stress can be considerable. For a cell whose internal osmotic concentration equals that of seawater, the entry of water upon immersion in freshwater creates a pressure of 22.4 atmospheres (22.4 times normal atmospheric pressure, about the same pressure experienced by a diver at a depth of half a kilometre). Conversely, freshwater organisms may die if exposed to hyperosmotic stresses by immersion in seawater. Extreme hyperosmotic stresses can develop in lakes and ponds where salts dissolved from the surrounding rocks become concentrated by the evaporation of water. The Dead Sea is one such place and has a salt concentration of 28 per cent.

As well as regulating their water contents, organisms also regulate the substances that are dissolved in their internal water. When sodium chloride dissolves in water, it splits into a sodium ion (Na^+) and a chloride ion (Cl^-). These, together with potassium ions (K^+), are the main inorganic ions found in the fluids of organisms. Lower concentrations of calcium (Ca^{2+}), magnesium (Mg^{2+}), sulphate (SO_4^{2-}), phosphate (PO_4^{3-}) and bicarbonate (HCO_3^-) ions are also found. High or low concentrations of these various ions may stress the organism.

Most organisms release energy from sugars such as glucose by utilising oxygen from the atmosphere to oxidise them (they 'burn' them via aerobic respiration to release the energy stored within their chemical structure). For these aerobic organisms, low oxygen concentrations in their environment may be stressful. Low levels of oxygen are found in environments where the oxygen has been used up by biological activity (such as in the centre of a compost heap or in the mud of an estuary) or in sites where there is limited access to oxygen (such as in the centre of the intestine of a cow). Some organisms, however, can survive temporary or permanent exposure to conditions where oxygen is absent, or at low concentrations, by respiring anaerobically (in the absence of oxygen). Most processes of anaerobic respiration use food much less efficiently than aerobic respiration. An anaerobic organism thus generally gets much less energy from a given amount of food than does an aerobic organism. This may limit the rate at which they can grow and reproduce.

Therefore, the main physical stresses on organisms are temperature extremes, pressure, desiccation, acidity or alkalinity, osmotic and ionic stress and low oxygen levels. Others include toxic chemicals and radiation (particularly, ultraviolet radiation).

The necessities of life

There are only three essentials for sustaining life – a source of energy, water and a range of conditions that the organism can tolerate (i.e. conditions that lie within the life box of the organism).

Living organisms must burn fuel for metabolism which drives growth, reproduction and the maintenance of the structure and integrity of their bodies. Animals and some microbes gain this fuel from the consumption or decomposition of other organisms (they are heterotrophs, 'feeding on others'). Plants and some other microbes are autotrophs ('self-feeders'), which means they can utilise energy to convert inorganic materials (such as water and carbon dioxide) to organic materials (such as those that make up the bodies of living organisms). There are two main sources of energy. Plants and plant-like microbes utilise the energy of sunlight via the process of photosynthesis (phototrophs, 'light-feeders'). Photosynthesis converts water and carbon dioxide, from the atmosphere, into glucose, an organic sugar. The sugar then acts as the fuel for the cell's metabolism. This is achieved most efficiently by breaking down the sugar in the presence, and with the involvement, of oxygen (oxidising or 'burning' the sugar). Some microbes can derive their energy from the oxidation of chemicals (chemotrophs, 'chemical feeders'). Sulphur bacteria get their energy by oxidising sulphur to sulphur dioxide, which then dissolves in water to form sulphurous acid and then sulphuric acid. Other bacteria perform similar tricks with iron, ammonia or hydrogen.

The second essential for life is water, or at least periodic access to water. Water is part of the structure of many of the molecules that make up living organisms. It also provides the medium in which the chemical reactions of living organisms take place. Without water, there is no metabolism and hence no life. However, as we will see, cryptobiotes can suspend metabolism, and perhaps life, and survive periods of complete water loss. Where there is liquid water and a source of energy which an organism can utilise, there will be life.

The forms of life
Let us look at some of the different types of organisms which we will encounter in this book. The forms of life are tremendously varied. Scientists have attempted to make sense of this variety by devising various schemes for arranging organisms into groups on the basis of their similarities and differences. This process, known as classification, has a long history. Linnaeus (1707–1778), a Swedish naturalist, was the first to attempt such a classification, dividing living things into animals and plants. Linnaeus was, however, limited to organisms he could see with his naked eye. The invention of the microscope and the discovery of the great variety of microorganisms made things much more complicated. Later schemes used various lines of evidence (such as morphology, embryology, geographical distribution, physiology and the fossil record) to develop classifications that attempted to express the evolutionary relationships between organisms.

One of the most important divisions of life is into organisms which consist of a single cell (unicellular) and those which consist of many cells (multicellular). Most microorganisms are unicellular. Plants, animals, fungi, slime moulds and some algae are multicellular. Another important division is based on the types of cells that make up the organism. Eukaryotic cells have their genetic material contained within a nucleus which is bounded by a nuclear membrane. The remaining part of the cell consists of the cytoplasm, which contains membrane-bound organelles of various types and functions. Prokaryotic cells are much simpler. The genetic material of a prokaryotic cell is concentrated into a particular region, but there is no membrane separating this from the rest of the cell and hence no nucleus. There are also no organelles in prokaryotic cells. The unicellular protists and yeasts are eukaryotes, as are most multicellular organisms. The remaining unicellular microorganisms are prokaryotes.

These classification schemes, which were based largely on the criteria of morphology and lifestyle, culminated in the five-kingdom model (proposed by Robert Whittaker of Cornell University in 1969) which divided life into five major groups or kingdoms (Figure 1.5): animals, plants, fungi, protists and bacteria. Animals are multicellular eukaryotes which rely on other organisms as a source of organic molecules (usually by eating them). There are 30 or so different phyla of animals (representing different body plans), ranging in complexity from sponges, jellyfish and worms through to birds, mammals and humans. You'll find descriptions of the main groups mentioned in the text in the glossary at the end of this book. Plants are multicellular eukaryotes which produce their own organic materials via the process of photosynthesis using energy from sunlight and which live mainly on land. The simplest plants (such as liverworts and bryophytes or mosses) lack roots or well-developed vascular tissues (tubes which transport water and nutrients). The simplest vascular plants (such as ferns and horsetails) do not produce seeds but reproduce via spores. Gymnosperms (conifers and cycads) produce naked seeds, unprotected by a seed coat, whereas angiosperms (flowering plants) have protected seeds.

Fungi are eukaryotes which do not photosynthesise but which do not ingest food either, as do animals, but absorb it after secreting

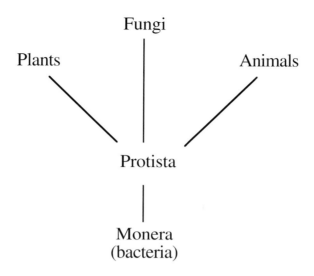

FIGURE 1.5 The five-kingdom classification of life.

enzymes to the outside of their bodies. They are mainly multicellular, but include the unicellular yeasts. The remaining unicellular eukaryotes are classified as protists along with some multicellular eukaryotes which do not fit in with the animals, plants or fungi. Protists are thus a bit of a grab bag of organisms (with some 30 to 40 phyla included) which are perhaps not that closely related. The protists include animal-like (protozoa), plant-like (algae) and fungus-like organisms (the multicellular slime moulds). Seaweeds are multicellular algae.

In the five-kingdom model, all prokaryotic organisms were grouped together as bacteria (Monera). More recent studies have indicated that there is at least as much diversity, and probably more, among prokaryotes as there is among eukaryotes and that there are several kingdoms of prokaryotes. Bacteria include organisms with a great variety of lifestyles including heterotrophs (which feed on organic molecules) and autotrophs (which make their own). The autotrophs include photoautotrophs (which photosynthesise, such as cyanobacteria) and chemoautotrophs (which obtain energy by oxidising various inorganic substances). Most are unicellular, although some form colonies or

aggregations and some even show a primitive multicellular organisation in which there is a division of labour between two or more specialised types of cell (some cyanobacteria).

The five-kingdom model clearly has trouble classifying all the
organisms that we can observe on Earth. This has led to additional
kingdoms being proposed and to groups of organisms being transferred
from one kingdom to another (such as including some multicellular
algae along with the plants). Some organisms, however, don't seem to
fit in with these schemes. Viruses consist of genetic material (DNA or
RNA) enclosed within a protein coat, and are not cells at all. They can
only reproduce by infecting the cell of another organism and using the
host cell's machinery to produce copies of the virus. Perhaps viruses
cannot even be considered to be alive, since they cannot reproduce
without the aid of a host cell. Lichens can be commonly observed
growing on the surface of rocks, walls and tombstones. They turn out
to be not a separate type of organism but a close association (symbiosis)
between two different types of organisms: a fungus and an alga or cyanobacterium.

The main problem with traditional models of classification is that
the criteria they use are, at least to some extent, subjective. The growth
in the methods of molecular biology has allowed the development of
molecular techniques that produce classifications based on more
objective criteria. These techniques compare the sequences of subunits (amino acids or nucleotides) that make up the structure of particular proteins and nucleic acids which are found throughout the
organisms to be classified. Mutations produce changes in the
sequences of these molecules which accumulate over time. Two organisms that have similar sequences are thus closely related and those
with different sequences are distantly related, with the degree of difference associated with how far back in evolutionary time they parted
company. These techniques have revealed a very different picture from
that based on morphological and physiological criteria (Figure 1.6). The
diversity of prokaryotes has been unveiled. The bacteria have been
shown to consist of two domains (each containing a number of king-

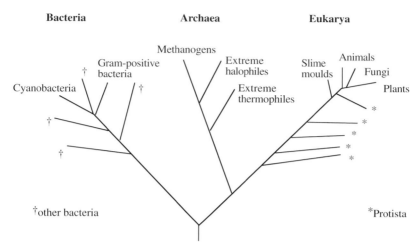

Bacteria   Archaea   Eukarya

FIGURE 1.6   A simplified tree of life based on studies which use molecular systematic data. Life is divided into three domains: Bacteria, Archaea and Eukarya. The branches represent kingdoms within these domains. Only kingdoms of organisms mentioned in this book are labelled. The protistans, which were grouped together in the five-kingdom classification, are recognised as separate kingdoms. The branches of the Archaea represent functional, rather than phylogenetic, groupings.

doms): the bacteria and the archaea. These are more different from one another than is a toadstool from a whale. The third domain, the eukarya, includes all the eukaryotic organisms. The archaea are associated particularly with extreme environments. They include three main functional groups: the methanogens (which produce methane gas as an end product of their metabolism), extreme halophiles (which live in very salty conditions) and extreme thermophiles (which live in high temperatures, such as hot springs and deep-sea hydrothermal vents). The archaea were once considered to be a form of bacteria (archaeabacteria) but they are now recognised as a distinct group. I may, however, in places still refer to them as bacteria – for which may the archaeal biologists forgive me!

Another way of grouping organisms is on the basis of their mode of life. Primary producers manufacture sugars and other organic compounds using the energy from sunlight (plants, algae, cyanobacteria and other photosynthetic bacteria) or from the oxidation of various inorganic substances (some groups of bacteria). Consumers (animals, fungi, some bacteria and protists) satisfy their requirements for organic molecules and energy by eating (or otherwise utilising the products of) the primary producers (or other consumers). Some consumers are decomposers (fungi, bacteria, some animals), deriving their food from dead organisms or organic waste such as faeces and fallen leaves, and some are parasites (or other types of symbionts), living in close association (within or on) other organisms.

Size matters

The size of an organism to a large extent determines how much of a problem it has with an environmental stress. Imagine: you are late for dinner. This is not the first time, your husband or wife is annoyed with you and, instead of putting your dinner back in the oven, has left it out on the table. If you're only 10 minutes late, the potatoes may still be hot but the peas have gone cold. Ten more minutes and the potatoes have gone cold too. How is it that the peas cool quicker than the potatoes? The answer is, of course, that size matters. Small things lose (or gain) heat faster than large things. This is because objects lose heat through their surface and the rate at which heat is lost from an object depends on the area of its surface compared with its volume. A pea has a larger surface area relative to its volume than does a potato and therefore loses its heat faster. A small object also has less heat to lose than a large object. Small organisms thus have more problems in regulating their temperature than do large organisms.

What is true for heat is also true for substances that pass through the surface to enter or leave an organism. The rate at which oxygen can pass (diffuse) through the surface of a small organism may be sufficient to supply it with the oxygen it needs. For large animals, such as humans, diffusion across the surface is not sufficient to supply their requirements. This problem is solved in humans, and most other terrestrial vertebrates, by the development of lungs. The lungs consist of a branching network of air passages (tracheae, bronchi, bronchioles) which end in air pockets (alveoli). This provides a large surface area across which the uptake of oxygen can occur. The total surface area of the lungs of humans is 100 square metres, while that of the skin is two square metres. The area available for oxygen uptake through the lungs is thus 50 times greater than would be available if oxygen uptake occurred via the skin.

While small organisms have much less of a problem with oxygen uptake than do large organisms, the reverse is true when it comes to desiccation. When exposed to desiccation, a small organism that loses water through its surface will dehydrate at a much faster rate than a large organism, because of its greater surface area relative to its volume. Desiccation is thus likely to be much more of a problem for a small organism than for a large organism. The organism cannot prevent desiccation by becoming completely impermeable to water – if the organism was impermeable to water, it would also not be able to breathe. There are no biological structures that are impermeable to water but permeable to oxygen. If water loss from the body is reduced by an impermeable skin or cuticle, there need to be openings, such as our noses or the pores in an insect's cuticle (spiracles) or in the surface of a plant (stomata), which let the organism breathe.

EXTREME BIOLOGY

In addition to physical stresses, organisms also face a number of biological stresses. These include competition with other organisms, predation, diseases and parasites, and the availability of food. Organisms that are able to survive or thrive under conditions which other organisms cannot are thus at an advantage. They can avoid competition by utilising habitats and food sources that are not available to others. Predators, parasites and diseases, which might otherwise affect them, may also be absent in their extreme homelands. In order to survive their unusual living conditions, extreme organisms have had to solve the problems and challenges posed by the stresses they experience. Extreme organisms take us to the outer edge of biology, an extreme biology, giving us new insights into the nature of life.

In the next chapter of this book, I will look at the different types of extreme environments found on Earth. The next three chapters deal with how some organisms cope with the main types of physical stress they experience in extreme environments: desiccation (Chapter 3), heat (Chapter 4), cold (Chapter 5), and pressure, pH extremes, osmotic stress, lack of oxygen, radiation and toxins (Chapter 6). In chapter 7, I will consider how studies on extreme organisms may help us in our search for extraterrestrial life and I will look at some general features of extreme biology in Chapter 8.

2   Be it ever so humble . . .

. . . there's no place like home (in the words of the poem by John Howard Payne). Although the environments inhabited by some organisms seem humble to us, to these organisms they are 'home'. Extreme habitats abound on Earth. They seem extreme to us because of high temperatures (hot springs, hydrothermal vents, hot deserts), low temperatures (polar regions, alpine environments, winter temperate environments, cold deserts), lack of water (deserts), high pressures (ocean depths), acidic or alkaline conditions (acid mine waste, the stomach, soda lakes), high salt concentrations (salt lakes) and lack of oxygen (decomposing organic material, estuarine muds, vertebrate intestine). Other extreme situations include toxic chemicals and exposure to ultraviolet radiation. Many extreme habitats challenge the organisms that live there with some combination of these stresses. Let us look at the characteristics of some of these extreme habitats and the adaptations which enable organisms to call them home.

DESERTS

Not all deserts are hot and sandy. A desert is defined by its lack of water rather than its temperature. This does not mean that there is no water but that rainfalls, and other inputs of water, are irregular and infrequent. Deserts cover one-third of the Earth's land surface (Figure 2.1). This includes semi-arid areas (annual precipitation of less than 600 millimetres), arid areas (less than 200 millimetres) and hyper-arid areas (less than 25 millimetres). Deserts form in the leeward side of mountain ranges, in inland areas which are remote from oceans and where dry stable air masses form, resisting convective currents which would bring rain. Most of the world's hot deserts (subtropical deserts such as the Australian and Arabian deserts and the Sahara Desert) are located

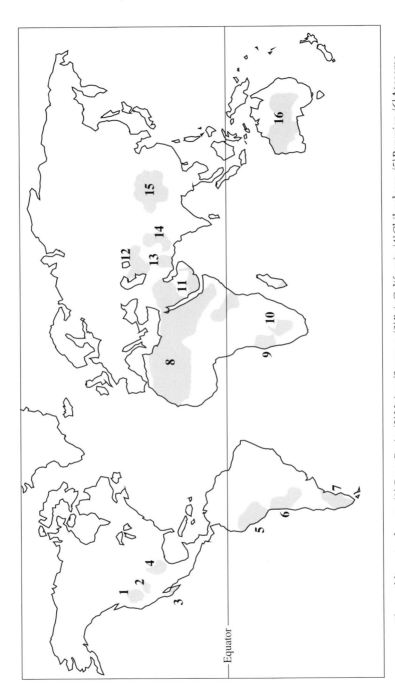

FIGURE 2.1 The world's major deserts. (1) Great Basin; (2) Mojave/Sonoran; (3) Baja California; (4) Chihuahuan; (5) Peruvian; (6) Atacama (Chile); (7) Patagonian; (8) Sahara; (9) Namib; (10) Kalahari; (11) Arabian; (12) Turkestan; (13) Iranian; (14) Thar; (15) Gobi; (16) Australian. Drawing by Ken Miller.

in the area between 25° to 35° north and south of the equator, with dry air tending to be trapped between major wind belts and storm systems. Cool coastal deserts (such as the Namib Desert, the Peruvian–Chilean deserts and the desert of Baja California in Mexico) form in subtropical areas where there are cold sea currents from polar regions. The moist, cold air formed by contact with the cold ocean current moves inland as a thin layer beneath the hot and dry tropical air. This does not produce rainfall, but results in condensation at night that the organisms of these deserts use as a source of moisture. Temperate or cold deserts form in the rain shadow of mountain ranges or in areas which are at a great distance from oceans. The Gobi Desert of central Asia is an example, whose elevation and distance from the coast results in extremely dry and cold conditions. The polar deserts of Antarctica are rather different from other deserts since there is no shortage of water in Antarctica – it is thought to hold 90 per cent of the world's store of freshwater, with 99 per cent of the surface of Antarctica being covered by ice. Water as ice or snow is, however, not available to organisms and the small areas of ice-free land in Antarctica are largely dry. The Dry Valleys of Victoria Land (also called the Ross Desert) are the most extensive areas of ice-free land in Antarctica. They are extremely dry and parts of them are thought not to have received any rainfall, or other precipitation, for at least the last two million years. Deserts result from low rainfall but they are also places of extreme temperatures and of rapid changes in temperature. In the Sahara Desert, air temperatures can regularly reach in excess of 50 °C and ground temperatures can be as high as 70 °C, while in the Gobi Desert air temperatures are as low as −40 °C at night.

When it does rain in a desert, the rain can be very heavy. This causes floods, but the water rapidly drains away or sinks into the ground. Water is thus only briefly available to desert organisms, unless they can access areas where water accumulates and persists for longer periods of time. Desert organisms have various mechanisms for coping with infrequent and unpredictable access to water. If there is no moisture available at all, however, there is no life. As well as limited access to water, desert organisms have the problem of hanging on to the water they have. This is partly because the air is very dry, but high temperatures and wind may also result in high rates of water loss from their bodies. They either have to survive the water loss or limit it so that they can retain sufficient water to last until it becomes available again.

The main stresses faced by desert organisms are thus infrequent and unpredictable access to water, high rates of water loss, extreme temperatures and large temperature ranges. Other problems may include precarious environments due to periodic floods, high winds and unstable ground, exposure to solar radiation and limited access to food or nutrients.

Living in the desert

Organisms show two broad responses to the extreme conditions that they face in deserts. They have adaptations which enable them to function in the face of low water availability and high temperatures or they escape, retreat from or avoid the harsh conditions, restricting their activity or growth to periods that are more favourable. This corresponds to the capacity adaptations (adaptations which allow the organism to operate under extreme conditions) and resistance adaptations (adaptations which allow the organism to avoid or survive the stress until the conditions favourable for its growth return) that were described in Chapter 1. Many organisms show a mixture of these two responses. Which response is favoured by the organism depends on the particular demands of its habitat, its evolutionary history and factors such as its size and degree of mobility.

Capacity adaptation with respect to low water availability involves the ability to access the sources of water which are available, the ability to restrict water loss from the organism and the ability to store water. The response to high temperatures involves the organism's ability to avoid gaining heat and the ability to lose heat. Large organisms (like a camel) have a greater ability to store water than do small organisms (like a bacterium) and they have less of a problem with water loss since they have a smaller surface area in relation to their volume. Problems of gaining heat are also diminished, for a similar reason, but, conversely, they are less able to lose heat. The suite of adaptations which the organism has evolved also depends on the complexity of the organism. Mammals have a more sophisticated suite of adaptations than do bacteria. Large animals can move greater distances than small animals, while plants and microorganisms may have no or little ability to move. Animals can have behavioural adaptations in response to the harsh conditions – for example, migrating to find water sources – while plants and microorganisms have no, or a restricted, ability to respond in this way.

Resistance adaptation involves escaping from or avoiding the harsh conditions. This operates on various time scales, ranging from avoiding activity during the hottest part of the day to being active only at night through to lying dormant for months or years until it rains and conditions favourable for growth return. Cryptobiosis is the ultimate resistance adaptation, with the organism capable of surviving in an ametabolic dormant state for many years. Again, the kind of adaptations which the organism has evolved depends on its evolutionary history. Resistance adaptations which involve behavioural responses, such as burrowing into the sand, are found in animals. Plants and microorganisms have a greater tendency to dormancy, particularly in the formation of seeds and spores which lie dormant until the return of favourable conditions.

Desert mammals

Mammals are the most complex of organisms and have evolved sophisticated behavioural and physiological responses to the demands of their environments. Their size and complexity does, however, impose some restrictions on the sorts of adaptations they have been able to develop and, for example, they are not capable of cryptobiosis. The camel is perhaps the most famous of desert mammals so let us look at why it is such a successful desert inhabitant.

There are two species of camels: the dromedary (one-humped or Arabian) camel and the Bactrian (two-humped) camel. Dromedaries are associated with hot, dry, flat deserts (subtropical deserts such as the Sahara Desert and Arabian Desert) while Bactrian camels live in mountainous, rocky regions (cold temperate deserts, such as the Gobi Desert). The two species coexist in Turkey, Afghanistan and Turkmenistan where they interbreed, indicating that they are closely related. The following discussion is focussed on the dromedary, which will be referred to simply as the camel. Dromedaries are entirely domesticated and so aspects of their biology are determined by their relationship to humans.

Camels are herbivores, browsing and grazing on desert vegetation. The desert vegetation is sparse because of the low rainfall. Annual plants, which provide lush grazing, appear only after rainfall and thus do not provide a reliable food source. Camels rely on hardy perennials which provide a more permanent food source that is adapted to the low rainfall. The main features of this food for the camel are that it is sparse, tough and salty, with small, curled leaves which may be reduced to thorns. Camels browse over a large area, often covering distances of 50 kilometres per day in their search for food. They browse sparsely, taking a bite from one plant and then a bite from another. This avoids overbrowsing the vegetation and allows it to recover; camels' use of their food source is thus sustainable. They eat a variety of plants and can eat plants which other animals cannot. Their ability to go for long periods without water means that they can feed on areas remote from water sources. The leathery lips and tongue of the camel allow it to contend with the tough vegetation and it can even eat the thorns themselves. The camel's physiology allows it to cope with the salt intake from its food and from drinking salty water; indeed, they need the salt and, in the absence of natural sources of salt, need to be given a daily ration of as much as 45–60 grams.

During the cooler months of the year, camels can rely entirely on their food plants for their intake of water, which on average consist of 50–60 per cent water during the cool season. Even when temperatures reach 30–35 °C, camels can go for as long as 15 days without drinking. It is only when temperatures exceed 40 °C that they need to drink at short and regular intervals. When they do drink, camels have a tremendous capacity to take in water. A camel which is dehydrated, after several days without water under hot conditions, may drink up to 200 litres of water over several hours and as much as 130 litres (about a bath full) in a few minutes during its first drinking session. This water is rapidly transferred to the camel's bloodstream, where it is available to rehydrate the tissues. This rapid dilution of the blood by an influx of water would be fatal for most animals since the resulting osmotic stress would cause the red blood cells to rupture. The camel's red blood cells are, however, unaffected. The ability to drink a large quantity of water in a short time also means that it can spend a minimum amount of time in overgrazed areas at water sources. In dehydrating conditions, the water content of the milk of lactating camels is actually increased, to ensure the survival of the young.

Camels can survive severe dehydration resulting in the loss of 20–25 per cent of their body weight (30 per cent of total body water). In contrast, humans die if they lose water equivalent to about 12 per cent of body weight. Most of this water is lost from the camel's gut and the spaces between the cells. Relatively little water is lost from the cells of the camel during dehydration. A dehydrated camel develops a hollow in its side behind the ribs. Nomads can judge to within 10 litres how much water a camel needs to drink from the shape of the hollow. No other mammal can survive without water for weeks on end and still remain active, a feat which makes camels invaluable to humans in arid areas. As well as surviving high levels of dehydration, camels have a number of mechanisms for conserving water within their bodies. The large body size of the camel means that the loss of water by evaporation through the skin is less than it is for smaller animals. Camels reduce their reliance on sweating to cool down at high temperatures. They produce a concentrated urine (high in salts and urea but low in water), relatively dry faeces and breathe slowly, reducing water loss in the breath. The requirement for a high salt intake is driven by the production of a concentrated urine. The nasal passages have large cavities which act to moisten dry inspired air and to recover water from expired gases.

There is no evidence that camels can store water anywhere in their bodies. Pliny the Elder (c. AD 23–79) was the first to suggest that camels store water in their stomach. However, there is no anatomical or physiological evidence that supports the suggestion that camels can store water in their stomach, or any other part of their body. Water is not stored in the hump. The hump consists mainly of fat and acts as a food store. The metabolism of fat does produce water, but this is exceeded by the amount of water lost in the gases expired during the respiration necessary to gain the oxygen involved in fat metabolism. Fat is an excellent insulator and if it was evenly distributed under the surface of the skin the camel would be unable to lose heat efficiently through its surface. Concentrating fat reserves in the hump (and around the kidneys) allows heat to be lost unimpeded over the remainder of the skin.

Most mammals control their internal temperatures within fairly tight limits. They lose heat by sweating, panting and by exposing areas of skin to the air. These mechanisms also result in water loss. Camels reduce the water loss by allowing their internal temperatures to fluctuate more than do other mammals. The difference between their internal temperature early in the morning and in the afternoon can be as much as 6 °C. By tolerating the increase in internal temperature, the camel conserves the water which would be lost through sweating and other mechanisms involved in losing heat. It is estimated that camels save as much as 5 litres of water per day by allowing their internal temperature to fluctuate in this way. Camels reduce heat gain by having a light-coloured skin and a thin woolly coat which shields the body from the sun. They face the sun, thus reducing the area exposed to heat radiation and the fat in their hump insulates the more exposed back. Their tall body shape and long legs allow heat to be lost from the hairless, poorly insulated lower half of the body and raises the body above the ground to where temperatures are cooler. The mechanisms which enable camels to survive in the desert are summarised in Figure 2.2.

Other large grazing mammals of subtropical deserts include antelopes, gazelles, zebu cattle and wild asses, and, in Australia, kangaroos

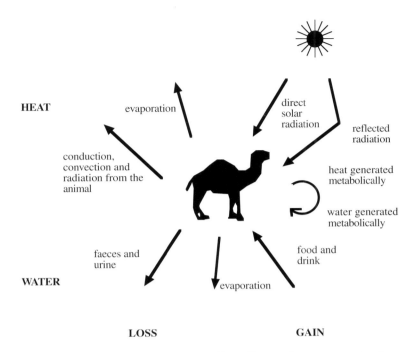

HEAT

evaporation

direct
solar
radiation

reflected
radiation

conduction,
convection and
radiation from the
animal

heat generated
metabolically

water generated
metabolically

faeces and
urine

food and
drink

WATER

evaporation

LOSS   **GAIN**

FIGURE 2.2 The major routes for the gain and loss of heat and water in an animal
such as a camel.

and wallabies. These share some of the adaptations to the desert environment shown by camels, although none show the same degree of
resistance. Mammalian desert carnivores include foxes, jackals,
hyenas, coyotes, small cats, badgers, skunks, ferrets, some Australian
carnivorous marsupials and the dingo. Only foxes are found in
extremely arid areas. These carnivores obtain most, if not all, of their
water requirements from consuming the bodies of their prey.

Small desert mammals can avoid the worst stresses of heat and desiccation by retreating during the day, or during the hottest part of the
day, into burrows or rock crevices. Most of these are rodents, no bigger
than the size of a rat. Not only is the temperature within the burrow
lower than that outside during the day, but the humidity is two to five
times higher, due in part to the breath of the animal. By avoiding the

heat of the day, desert rodents escape the temperature regulation problems posed by exposure to the sun. They do not sweat since their small body size, and hence relatively large surface area, means they cannot survive the loss of water. A rodent may, however, find itself trapped outside during the day – due, perhaps, to being driven from its burrow by a predator. Kangaroo-rats have an emergency temperature regulation system which enables them to survive in such situations. They produce copious saliva which wets the fur of the chin and the throat, producing cooling by evaporation. Normally, water loss is confined to the breath and this loss is reduced by cooling the air as it leaves the nose, so that it carries less water. Other water conservation measures include the production of a concentrated urine and dry faeces. Many desert rodents rely entirely on their food as a source of water. Egyptian jerboas and American kangaroo-rats can live indefinitely on a dry diet of plant seeds. Metabolic water, produced as a product of the oxidation of food, can be a major source of water for animals in arid areas.

Other desert vertebrates

Desert birds can travel large distances to obtain water. Their water supply problems become more acute during the breeding season when they must also satisfy the needs of their nestlings. The African sand grouse often nests 40 kilometres away from standing water. The male bird ferries water for his chicks by soaking his belly feathers. These have a unique structure which enables them to act like a sponge, soaking up and holding the water. The American 'Road Runner' bird produces a liquid from its stomach which trickles down the back of its throat and is drunk by its nestlings. The feathers of birds not only enable them to fly but are also very good insulation, protecting the bird from the heat of the sun. If they need to lose heat, they flutter their throats, creating a cooling flow of air through the moist inside of the mouth. Ostriches can raise their sparsely distributed dorsal feathers, allowing a cooling flow of air over a large area of their skin surface.

Desert reptiles, such as snakes, lizards, geckos and tortoises, like small mammals, avoid the hottest part of the day by burrowing into the ground or by sheltering in rock crevices. They have to perform a delicate balancing act apropos their internal temperature. The temperature needs to be high enough to allow the animal to be active, but, if it is too high, the animal will die. Reptiles have a very limited ability to generate their own heat, in contrast to mammals and birds, and rely on the heat of the sun to raise their temperature sufficiently for activity to occur. It is often cold at night in the desert and few reptiles are nocturnal. In the morning, they bask in the sun, absorbing the heat and raising their internal temperature to a level where they can become active. As the heat of the day increases, they retreat to their shelters. During the late afternoon and early evening, temperatures become suitable for activity again. Reptiles can extend their periods of activity by adaptations which enable them to gain or lose heat faster when they need to. Some Namib dune lizards, when they emerge in the morning, press their bodies against the warm sand to gain heat from the surface. As the temperature rises, they adopt a stilt-like posture which raises the body above the surface and periodically raise diagonally opposed limbs, allowing more heat loss.

Desert amphibians have particular problems with water availability, since they need to lay their eggs in water. They are consequently restricted to areas of deserts where a suitable body of water is at least periodically available. Desert ponds form after heavy rainfall and rapidly dry up. The rainfalls which supply sufficient water for amphibians to breed may be separated by long dry periods. One of the best known desert amphibians is the spadefoot toad of the Sonoran Desert of North America. The toads go into a dormant state during dry periods. They burrow nearly a metre into the ground to find relatively moist conditions where they can remain for as long as nine months. Some amphibians are surrounded by a mud cocoon while others produce a cocoon of dry dead skin which prevents rapid water loss. The animals also accumulate urea which raises the osmotic concentration of their tissues and results in a net flow of water into their bodies from the surrounding soil. They can also tolerate dehydration of the tissues and store water in the form of a dilute urine in the bladder. When it rains, the spadefoot toad emerges and breeds. The eggs develop rapidly and the pond becomes full of tadpoles. The pond may also become colonised by algae and fairy shrimps (small crustaceans). If fairy shrimps are present, two types of tadpoles are formed. Herbivorous tadpoles feed on the algae while larger carnivorous tadpoles feed not only on the shrimps but also on their herbivorous colleagues. The formation of these two types of tadpole is an insurance policy for the toad species. If it rains again, the pond will become muddy and the carnivorous tadpoles will not grow well since it is difficult for them to see their prey. The herbivorous tadpoles will continue to feed on the algae and may mature in large numbers. If it fails to rain, it is important that at least some individuals complete their development. The carnivorous and cannibalistic tadpoles develop quickly by consuming their siblings. They compete for the rapidly disappearing puddles of water and a few survive to breed after the next rainfall.

Desert invertebrates

There are three broad groups of desert invertebrates: those permanently inhabiting the soil (dominated by nematodes, mites and springtails), those active on the surface for at least part of the time (centipedes, spiders, scorpions and a wide variety of insects) and those associated with temporary bodies of water (shrimps, crabs, nematodes and some insects). Other less prominent desert invertebrates include snails, woodlice, earthworms and millipedes.

Soil invertebrates feed on organic material that is buried by the action of wind and rain, on the microorganisms associated with its decomposition or on other soil invertebrates. Plant material is also buried by the activities of animals which are active on the surface, such as termites, ants and other insects. The highest densities of nematodes are found close to desert vegetation and in the top 10 centimetres of the soil. While the soil and the proximity of vegetation provides some protection against high temperatures and high rates of water loss, desiccation is likely to be a major challenge to their survival.

Arthropods active on the surface of the ground still spend a major part of their time beneath the surface where they are protected against the harsh environment and may only emerge to forage, collect water and perhaps find a mate. Ants and termites are major components of the invertebrate fauna of deserts, with their biomass exceeding that of all other invertebrates combined and even that of vertebrates in some places. A range of ant species is found in deserts, particularly in the arid regions of Australia. They live in nests beneath the ground. Termites are found in deserts throughout the world. Their nests extend both above and below the ground, but, in the hottest deserts, the nests are almost entirely underground. A foraging area is established around the nest. Most desert termites have little resistance to water loss and only forage when the humidity is high. Ants are more tolerant of desiccation but still retreat to the nest when temperatures are high. The architecture of the elaborately constructed termite mounds creates a flow of air which ventilates and cools the nest, and the orientation of the mound to the sun means that they tend to only absorb heat when air temperatures are low.

A wide variety of other groups of insects feed on desert vegetation – among the most prominent are the grasshoppers and locusts. Desert locusts occur in two forms which for a long time were thought to be separate species. Their life cycle allows them to take advantage of the flush of desert vegetation that occurs following rain. During dry periods, they are pale in colour, solitary and breed once per year. When it rains, the resulting flush of vegetation allows many more of the young to survive than do during dry periods. These become crowded together on the dwindling vegetation as it shrinks during subsequent dry periods. The crowding stimulates them to convert to the darker gregarious form and they can produce huge swarms numbering in the millions. Low pressure weather systems stimulate them to launch into the air where they are carried in a weather pattern which is likely to produce more rain and hence vegetation and food. These swarms have enormous destructive power and are a threat to human agriculture since they devour all vegetation in their path. Eventually, the numbers dwindle and they resume the solitary form.

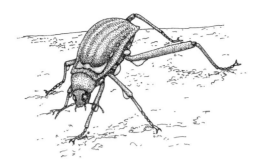

FIGURE 2.3 Namib beetle (tenebrionid) collecting moisture from fog – water droplets accumulate on its body. Drawing by Jo Ogier.

Beetles from the family Tenebrionidae are often thought to be the insects that are best adapted to desert life, with their numbers increasing in areas adverse to other forms of life. They can live on dry foods without any water and some are active during the hottest part of the day. They are extremely tolerant of dehydration. The osmotic concentration of their body fluids (haemolymph) is maintained during episodes of dehydration and rehydration, although the haemolymph volume changes, thus maintaining the physiological integrity of their cells. Tenebrionids have a very waxy cuticle, which reduces water loss across their surface. Their dorsal wing cases (elytra) are fused and form a cavity covering their backs. The respiratory openings (spiracles) open into this cavity and hence are not in direct contact with dry air. This reduces the amount of water lost during respiration. They reabsorb water from the rectum and produce very dry faeces. Namib tenebrionid beetles collect moisture from the fog which originates from the cold coastal currents offshore from the Namib Desert. When fogs occur at night, they migrate to the crest of a sand dune, where condensation is greatest. They stand with their heads down and face into the wind. Moisture from the fog condenses on their backs and trickles down to their mouths (Figure 2.3).

Insects and other arthropods are such successful inhabitants of deserts, despite their small size and hence relatively large surface area, because of their ability to restrict water loss. They do this by having a waxy cuticle, which is impermeable to water, and a respiratory system involving openings in the cuticle and air tubes (tracheae and tracheoles), which supplies sufficient oxygen to the tissues while restricting water loss. Water loss via the faeces is restricted by reabsorbing water through the rectum. Water is gained by the ingestion of food and collecting condensation and some species can absorb water from the air, even at relative humidities as low as 81 per cent. Water within the cells is maintained, with the volume of water in the body cavity decreasing during dehydration while maintaining its osmotic concentration. The most extreme conditions may be avoided by burrowing, by dormancy during adverse periods and by daily and seasonal patterns of activity and reproduction. Temperature regulation may also involve body colouration and behaviours, postures and orientations which allow the animal to gain or lose heat.

The invertebrate inhabitants of temporary desert ponds have to survive periods of extreme desiccation. Such problems are also faced by those inhabiting temporary bodies of water in other environments and will be considered later in this chapter.

Desert plants

Almost all desert areas support some sort of vegetation, with dunes and bare rock supporting the least. Plants face similar problems to those of animals in desert areas, particularly extreme temperatures and low, irregular and unpredictable water supply. In some ways, their water supply problems are worse than those faced by animals since they cannot rely on their food as a source of water. Plants produce the sugars they use to fuel their metabolism via photosynthesis. They need carbon dioxide from the air and water for photosynthesis. The carbon dioxide enters the plant via pores (stomata) in the leaves. If the stomata are open to allow the entry of air, they also allow the loss of water. This process is called transpiration. Plants can compensate for the water lost through transpiration by taking up water through the roots. In a desert, the air is very dry, resulting potentially in very high rates of transpiration. This, together with low water availability, means that desert plants have problems maintaining their water content.

Desert plants have solved their water problems in a variety of ways. The vast majority of desert plant species are ephemeral. One of the most spectacular sights in nature is the growth and blooming of ephemeral desert plants after a rainfall. Ephemeral annual plants lie dormant as seeds which only germinate after rain. They then grow, flower and set seed within just a few weeks so that they can complete their life cycle while water is still available. This is, of course, a resistance strategy with the plant lying dormant, in the form of seed, until favourable conditions return. The seeds may be dry enough for their metabolism to cease and to become anhydrobiotic. Germination of seed after rainfall could be a very risky strategy since there may not be sufficient water for the plant to complete its development and it may wither and die before setting seed again. Some desert plants have solved this problem by having a chemical in the seed which inhibits germination. If there is sufficient rainfall, the chemical is washed away and the seed germinates, but, if there is only a little water available, germination is prevented. Annuals (which die after setting seed) make up the majority of ephemeral plants. Ephemeral perennials survive as an underground bulb or corm. Ephemeral annuals are favoured in deserts with the greatest variation in rainfall. In the most arid areas of North American deserts, such as Death Valley, where rainfall is very variable, ephemeral annuals make up 96 per cent of plants. Where there is less variability in rainfall, perennials predominate.

Perennial plants survive in desert environments by gathering water efficiently and by adaptations which allow them to retain that water. Temperate trees and shrubs have as much tissue below ground, in the form of roots, as there is above ground. Desert plants have from two to six times as much tissue below ground as above, allowing their roots to gather water over a wide area. This means that the plants are widely spaced and that desert vegetation is typically sparse. Some desert trees and shrubs can send out lateral roots as far as 75 metres to reach water sources. Transpiration is reduced by decreasing the surface area through which water can be lost. The cuticle of desert plants is waxy so that water loss is restricted to open stomata. Most of the plant tissue is below ground, leaves are small and reduced to spines or are lost altogether, with the stem taking over the role of photosynthesis. Some plants only grow leaves after rain and shed them again as the soil dries out. Other desert plants have remarkable abilities to store water after rainfall. Cacti, yuccas and euphorbias are called succulents because their thick, spongy leaves and stems can absorb and store large quantities of water. The stems of some cacti are pleated into grooves which enable them to expand as the plants take up water after rain. The mature saguaro cactus is able to store up to eight tons of water and can survive up to two years without rainfall. Succulents also have the ability to store the carbon dioxide which is produced by their metabolism at night and reuse it during the day for photosynthesis. This recycling of carbon dioxide reduces the need to allow air into the plant and thus reduces the loss of water through transpiration.

Many of the adaptations of desert plants have more to do with protecting the plant against excessive heat than with preventing water loss. The leaves of desert plants and the pads and stems of succulents are orientated with their thin edge facing the sun, reducing heat gain. Reduction of leaves to form spines protects the plant against heat gain by reflecting solar radiation. The spines also create a still layer of air which reduces the rate of heat gain from the surrounding air and protect the plant against grazing animals.

Desert microorganisms

At first glance, many desert soils appear lifeless and sterile. In fact, they contain numerous microorganisms such as bacteria, algae, protozoa and fungi. There is probably no soil in the desert, or indeed anywhere on Earth, that is entirely sterile. The numbers of microorganisms are, of course, related to the availability of nutrients and of water. They are most numerous in the upper part of the soil and in association with the roots of plants. Large areas of deserts are covered by hardened surfaces called, variously, desert pavements, desert crusts and desert varnishes. These are inhabited by microorganisms, which are involved in their formation, and such soils are also referred to as cryptogamic crusts.

These crusts aid the survival of the microorganisms which help form them by modifying the habitat so that it is more stable and retains water better. Blue-green algae from the Atacama Desert of northern Chile produce gelatinous sheaths which trap sand grains and form cushion-like pellets which shade the small area of soil beneath and help them retain water.

Desert microorganisms can gather enough moisture from rainfall and/or condensation to support their growth, but numbers are drastically reduced when the soil dries out. Some can produce cysts, spores or mucilage which reduce the rate of water loss and enable them to survive anhydrobiotically. Normal growth forms, however, have little ability to prevent water loss. In spite of this, spore-forming bacteria make up a minority of the total number of bacterial species. In order to survive in dry soil, microorganisms must either be able to survive desiccation by anhydrobiosis or able to recolonise the area when water returns. We know much less about the abilities of microorganisms to survive dry conditions than we do about their survival at high temperatures (thermophiles) and high salt concentrations (halophiles).

Some microorganisms avoid harsh conditions by sheltering in favourable microhabitats, such as under a rock or pebble. Some even live within rocks (endolithic microorganisms), either in fissures or cracks (chasmoendolithic) or within the porous structure of the rock itself (cryptoendolithic). Photosynthetic bacteria (cyanobacteria) or algae form the basis of these microbial communities, with fungi, lichens and other bacteria associated with them. The rock must be transparent and the microbes close enough to its surface to allow light to reach them for photosynthesis, but the layer of overlying rock is sufficient to protect them against desiccation and other hazards.

TEMPORARY DESERTS AND TEMPORARY WATERS

Deserts are not the only places where organisms may have problems with desiccation. Have you cleaned out your gutters recently? If you have not done it for a while, you may well find things growing in them. I am writing this section while staying at Rothamsted Manor, which accommodates visiting scientists and students working at Rothamsted Experimental Station in Harpenden – the main agricultural research station in the UK. My room has a ladder leading onto the roof, providing a fire escape. The slate roof of the Manor is a temporary desert ecosystem. The tiles are covered with lichens and there is moss growing between them. Most of the Manor dates from the sixteenth century, although parts are from the twelfth century, and the lichens may well be several hundred years old. The mosses tend to grow in areas where water accumulates when it rains. Take a look around the outside of your house. Wherever water accumulates during rain, and where conditions are relatively undisturbed, you are likely to find mosses growing. Drier areas (like the surface of the roof tiles) may be colonised by lichens and cyanobacteria.

I took a scraping of the dry moss and, after soaking it in water overnight, examined it under a microscope. I could see several different kinds of organisms living in association with the moss. These included microscopic animals – rotifers, tardigrades and nematodes – as well as microorganisms such as protozoa, fungi and bacteria. There is a whole community of organisms living on the roof of the Manor. These observations repeat those of Antoni Van Leeuwenhoek, a seventeenth-century Dutch scientist, who was one of the first to observe living organisms using the recently invented microscope. His description of what he called 'animalcules' is given in Chapter 3.

The roof of the Manor is a temporary desert because, although it rains quite a lot during the year, the water rapidly drains away and the roof organisms have to survive periods of desiccation. There is little material to hold the water for long and the moss dries out. The roof community thus has to survive periods of extreme desiccation lasting several days or even weeks. They are also likely to be exposed to quite high temperatures. The air temperature here, while writing this, was 29 °C and the surface of the dark slate roof, which absorbs the heat of the sun, was likely to be much higher. Similar habitats provided by human activity include the tops and sides of walls and other structures, and 'container habitats' such as discarded tin cans, bottles, cooking pots and car tyres. Natural and artificial container habitats are particularly significant in the tropics where they often act as breeding grounds for mosquitoes which transmit diseases such as malaria and yellow fever.

There are many more natural situations where organisms are exposed to periods of desiccation, even though the rainfall of the general area they inhabit means it is far from being a desert – their microhabitat is a temporary desert. Some examples include the exposed surface of the soil and the edges of lakes, ponds and streams which desiccate when the water level falls. Small bodies of water may dry out completely. Organisms living on the surface of the aerial parts of plants (the bits above the ground, such as the bark of a tree or the canopy of a forest) and those living in plant tissue which dries out when the plant dies or sets seed are exposed to desiccation. Natural container habitats provided by plants include treeholes, the junctions between leaves and stems (for example, in bromeliads) and in flowers. Some of these periodically dry out.

Parasites living in the intestine of other animals need to infect new hosts to ensure their survival. They form eggs, larvae or cysts which pass out of the host when it defecates and initiate the free-living phase of their life cycle. Dung is also colonised by a variety of free-living organisms such as nematodes, earthworms, fly larvae, beetles, fungi and bacteria. Fly larvae must develop into adults before the dung desiccates or disintegrates and have very rapid rates of developments. Free-living nematodes and microorganisms and the free-living stages of parasites can tolerate the desiccation and remain within the dry dung until they are liberated by rainfall.

Although the area in which we live may seem far from being a desert, we can, if we look carefully, find many places which are temporary deserts where organisms will be exposed to desiccation and other extreme environmental stresses.

Temporary ponds and streams, which periodically dry up, are found in most parts of the world. These are inhabited by organisms that can survive the stresses involved in dry phases of varying durations, intervals and intensities. Desert ponds and streams provide some of the most extreme of these types of habitat. These form only after rainfall. Winds and hot, dry air produce rapid rates of evaporation meaning that ponds can be very short lived, often lasting only one or two weeks. Rain-filled rock pools which form in shallow depressions in rocks are common features of tropical and subtropical regions. These contain water for varying lengths of time, from a day up to several weeks. The more short-lived the pools, the greater are the demands on the organisms that live there.

Organisms have various ways by which they can exploit temporary bodies of water. They may be transient inhabitants, colonising the pond when it fills with water and leaving it when it dries. Many insects only live as larvae within these waters and rapidly complete their development to emerge as adults which fly away to seek more permanent waters. As we have seen, desert amphibians burrow into the mud as the pond dries and lie dormant until the next rainfall. Although we think of frogs and toads as being associated with moist environments, they can in fact successfully survive in periodically dry areas. In the western deserts of Australia, the numbers of frogs emerging from their burrows after rain is so large that they can disrupt rail travel – the rail tracks become slippery as thousands of frogs are crushed beneath the wheels of the trains.

Other animals also escape the threat of desiccation by burrowing. Some freshwater crayfish can burrow to depths of a metre or more to reach ground water. They remain active in their burrows by building chimneys which supply oxygen, but these can be plugged with mud to reduce water loss during dry weather. The water at the bottom of these crayfish burrows is used as a refuge for a variety of small invertebrates and a whole community of animals inhabits them. Other crustaceans burrow to avoid adverse conditions, as do some pond snails.

The African lungfish can survive when its river habitat dries. As the water level falls, it burrows into the mud of the river and secretes a cocoon of mucus (Figure 2.4). The cocoon is waterproofed by a layer of lipid and the fish lies folded on itself with its head next to a small

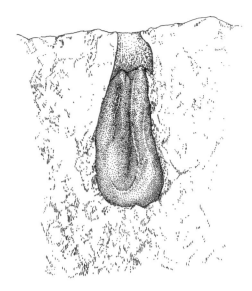

FIGURE 2.4 The African lungfish in its cocoon in a dry river bed. The lungfish is up to 2 metres long when fully grown. Drawing by Jo Ogier.

opening at the top of the cocoon. The fish's oxygen uptake reduces to 10 per cent of normal, its heartbeat slows, its tissues partly dehydrate, urine production ceases and the fish switches from producing ammonia to urea, which is less toxic and can accumulate in the blood and tissues without adverse effects. Part of the cocoon extends into the mouth of the fish and acts as a respiratory tube. The lungfish can survive in this dormant state for at least six months. When water returns, the lungfish becomes active again in a matter of minutes. The fish cannot rely on its gills to supply it with oxygen and it must swim to the surface in order to reach the air before it drowns.

Most invertebrate animals and all microorganisms are too small to escape desiccation by burrowing very far and their relatively large surface area in relation to their volume means that they cannot prevent water from being lost from their bodies. They survive by becoming dormant and entering into anhydrobiosis. Small crustaceans are some of the most obvious inhabitants of the most transient of temporary ponds. These include fairy and brine shrimps (Anostraca), tadpole shrimps (Notostraca), clam shrimps (Conchostraca), water fleas (Cladocera), seed shrimps (Ostracoda) and copepods. Some species survive anhydrobiotically as a juvenile stage, with a few forming protective cysts. Most species produce resistant eggs which lie dormant in the dried mud of the pond and hatch en masse when it rains. Anhydrobiosis thus not only allows the animal to survive the period of desiccation but also allows its life cycle to become synchronised with periods of favourable conditions. Only a few days after rain, these ponds can be swarming with millions of tiny shrimps which have emerged from eggs lying dormant in the dried mud.

Other invertebrates of temporary ponds survive periods of desiccation in a state of anhydrobiosis. These include nematodes, rotifers, tardigrades and some insect larvae. One of the most spectacular examples is the larva of a midge (a chironomid insect), *Polypedilium vanderplanki*. This inhabits shallow, exposed, rain-filled rock pools in the African 'kopjes' (isolated hillsides) of Nigeria and Uganda. At the start of the rainy season, these ponds may fill with water and dry out several times. Even during the dry season, it may rain and the ponds briefly fill with water. The larvae are thus exposed to cycles of desiccation and rehydration. The larvae can survive anhydrobiotically and can lose 99 per cent of their body water. This is the largest animal known to survive anhydrobiosis.

Microorganisms, such as protozoa, bacteria, fungi and algae, are found in abundance in temporary waters. Little is known of their adaptations to these habitats, although they may produce cysts (protozoa), spores (bacteria and fungi) or modified cells with thickened walls, mucilage sheaths and an accumulation of oils (algae) which enable them to survive anhydrobiotically.

SALT LAKES AND SODA LAKES

Freshwater never consists solely of water but contains, dissolved within it, other substances from the atmosphere or the soil. Lakes which are fed by water flowing over soils and rocks that contain many easily dissolved minerals can become quite salty. Where the lake is fed by streams and rivers but there is no outlet, evaporation may exceed the water flowing in and the minerals become trapped and concentrated. If the minerals are predominantly chlorides, mainly sodium chloride (common salt) and magnesium chloride, salt lakes are formed with the water being distinctly briny. Lakes containing more than 5 grams of minerals per litre of water are considered to be salt lakes. Seawater contains 35 grams per litre, but some salt lakes contain minerals at concentrations many times that of sea water. Where the minerals are predominantly carbonates and bicarbonates, particularly in volcanic areas, the water becomes alkaline and soapy, forming soda lakes.

The largest salt lake in the world is the Caspian Sea. This is fed by freshwater from the Volga River, but has no outlet and hence salts accumulate. Although the Caspian Sea is quite salty, it is dilute enough to support a fairly normal range of organisms. Some salt lakes are so salty that only very specialised organisms can live in them. The best known are the Dead Sea in the Middle East and the Great Salt Lake in the USA, but they are found in many parts of the world. If evaporation continues to exceed the flow of water into a salt lake, it will eventually dry up completely, leaving behind a salt flat consisting of a large expanse of brilliant white crystals. This has happened to many of the salt lakes of the Australian desert, but the largest salt flat is Salar de Uyuni high in the Andes of Bolivia. This forms a sea of salt 100 miles long and 85 miles wide. The salt flats may fill with water after heavy rain, but strong winds or high temperatures make them such efficient evaporation pans that they rapidly dry out again.

The Dead Sea is the most concentrated natural salt lake in the world and lies at the lowest point on the Earth's land surface. About one-third of its volume consists of minerals in solution, making it so dense that bathers find it almost impossible to sink or dive in it. The severe osmotic stress such high salt concentrations cause means that few organisms can live under these conditions. High levels of magnesium and calcium found in the Dead Sea appear to be a particular problem. The lake's food chain consists mainly of just two groups which can tolerate these conditions: an alga and several species of halobacteria (salt-tolerant archaea). At times, the lake develops a reddish tinge. This is due to blooms of an alga, *Dunaliella parva*, which, together with related species, is a common inhabitant of saline lakes and ponds. The halobacteria live off compounds produced by the algae. Several species of halobacteria, including *Halobacterium halobium*, have been isolated from the Dead Sea and are unique to it. How the algae and bacteria can tolerate such high salt concentrations will be considered in Chapter 6. Much of the water from the Jordan River, which is the main supply of freshwater to the lake, has been diverted and this often raises the salt concentration beyond that in which even these organisms can grow: the Dead Sea is dying.

The salt concentrations in some other salt lakes are low enough to permit the growth of some multicellular organisms. The most characteristic animal of salt lakes is the brine shrimp, *Artemia*. This is a crustacean belonging to the group known as fairy shrimps (Anostraca), which are common inhabitants of temporary ponds and saline waters. These feed on the halobacteria and algae and, since they have few competitors, can grow in enormous numbers under the right conditions. They are restricted to lakes and ponds which are too saline to support fish and other predators which would rapidly eat the brine shrimps. The fact that they are so palatable to fish was discovered in the 1950s by CC Sanders, a keen keeper of tropical fish. This led to an industry harvesting brine shrimps from the Great Salt Lake for sale as fish food. At its peak, in 1965, this industry harvested 77.2 tons each year of brine shrimp cysts, an indication of the enormous numbers growing in the lake. The numbers, and the industry, have since declined. Increased rainfall and snow in the mountains have caused the lake level to rise. The more dilute water has allowed the growth of waterboatmen, aquatic insects that feed on, and have reduced, the brine shrimp population. Brine shrimps have remarkable abilities to survive both high salt concentrations and desiccation, phenomena which will be explored in Chapters 3 and 6. Their survival abilities have led to the development of another, rather unusual, industry. Packets of dried cysts are marketed as instant pets called 'sea-monkeys' – just add water and nutrients and you have 'instant life'. Sea-monkeys have developed quite a following, as you will discover if you check them out on the Internet (try, for example, http://www.sea-monkeys.com/).

One of the best-known soda lakes is Lake Nakuru in the East African Rift Valley. This is famous because it supports large numbers of flamingos, at times numbering as many as 1.5 million. The minerals in the lake are mainly carbonates and bicarbonates which make the water alkaline. Organisms living in the lake have to survive the alkaline conditions and also wide variations in mineral concentrations as the lake level falls and rises due to evaporation or rain. The main photosynthetic organism is a cyanobacterium, *Spirulina platensis*. This is eaten by one species of copepod crustacean, one fish and by the lesser flamingo. Rotifers, waterboatmen and midge larvae are also found in the lake. *Spirulina* contains an unusually high proportion of protein. It forms a thick scum around the edge of soda lakes which is harvested in Mexico and in Chad by local people to make nutritious biscuits. It may be useful as both a source of protein and of animal feed. The lesser flamingo has filters in its beak which enable it to extract *Spirulina* from the water. The spacing of the filter in the beak of the greater flamingo is larger and they feed on the small invertebrates; thus, where the two flamingos occur together, they do not compete for food.

THE ANTARCTIC

Antarctica is a big place. It is the Earth's fifth largest continent, accounting for 10 per cent of its land surface and covering an area of some 14 million square kilometres. It is twice the size of Australia and half as big again as the USA. Surrounding it is the Southern Ocean which covers an area twice that of the Antarctic continent and which isolates Antarctica from other land masses. As might be imagined, while there are some common features to Antarctic environments, there is also a tremendous variety. We can recognise some broad distinctions, however, relevant to life in Antarctica. Antarctic marine environments are less extreme than terrestrial environments, due to

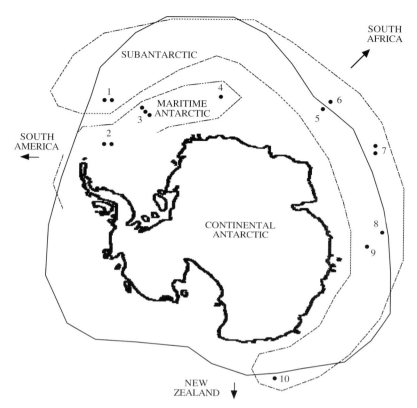

FIGURE 2.5 Map of the Antarctic region, showing the approximate position of the Antarctic Polar Front (solid line) and of the Subantarctic, Maritime Antarctic and Continental Antarctic regions. The dots indicate the approximate position of the major Subantarctic and Maritime Antarctic islands: (1) S. Georgia; (2) S. Orkney; (3) S. Sandwich; (4) Bouvet; (5) Marion; (6) Prince Edward; (7) Crozet; (8) Kerguelen; (9) Heard; (10) Macquarie.

the lack of exposure to desiccation and the buffering effects of seawater temperatures. The Antarctic Polar Front (Antarctic Convergence) is recognised as the northern boundary of the Southern Ocean (Figure 2.5). This is where the northerly flowing currents from the continent sink beneath the warmer currents which circulate around the Southern Ocean. This not only represents a change in seawater temperature but also in chemical composition and serves to isolate many Antarctic marine organisms. The marine habitat varies with latitude and depth and in its relation to sea ice.

Terrestrial Antarctic environments are much more extreme and life only occurs in sparse areas where the conditions are favourable for its survival. Three broad ecological zones are recognised in terrestrial Antarctic environments. The subantarctic consists of a scattered ring of small islands which lie within, or close to, the Antarctic Polar Front (Figure 2.5). These experience a wet, cool oceanic climate with temperatures above freezing for at least six months of the year and an annual precipitation of over 900 millimetres. The maritime Antarctic consists of the west coast of the northern part of the Antarctic Peninsula, its nearby islands and some island groups, such as the South Sandwich, South Orkney and South Shetland Islands, which lie in the Scotia Basin. This region has mean daily summer temperatures above 0 °C and winter temperatures rarely below −10 to −15 °C. The remaining part of the Antarctic, and its associated islands, is considered to be part of the continental Antarctic region. I will focus on life in this region, since this is where the most extreme environmental conditions are found.

The harshness of the climate of the continental Antarctic increases as you travel towards its centre. At the coast, summer temperatures exceed 0 °C for up to one month in summer, and winter means range from −5 °C to below −20 °C. On the ice plateau, which reaches elevations over 3000 metres, mean monthly temperatures are below −15 °C all year round, falling to below −30 °C in winter. Life is mainly associated with ice-free areas. These are in short supply with less than 1 per cent of the continent ever free of snow and ice. There are three types of bare ground. Some areas, notably the Dry Valleys of Victoria Land, are permanently free of snow and ice due to local climatic conditions. Other areas, mainly on the coast, have some snow cover in winter which melts during summer to provide liquid water. In places, exposed areas of rock project above the ice sheet (nunataks) and are associated with mountains which are isolated among the snow and ice.

Antarctic terrestrial organisms
When you say 'Antarctic animals', most people think of penguins and seals. These are, however, part of the community of marine organisms.

They breed on the continent, but return to the sea to feed and, in most cases, to escape the rigours of the Antarctic winter. The true animal residents of the continent, which live there all year round, are fully terrestrial and microscopic, with the largest (mites and springtails) being no larger than a pinhead (1–2 millimetres in length). The highest forms of terrestrial animal life are two species of midge, which are found in a few sheltered areas on the west coast of the Antarctic Peninsula and on the South Shetland Islands.

When you visit one of the ice-free areas of the continental Antarctic, the overwhelming impression is one of barren rock, gravel and scree. Yet, look carefully and, in a few favoured sites, there is life even here in one of the harshest environments on Earth. The ground may be tinged green by the growth of algae and black crusts on the soil surface indicate the presence of microorganisms. There may even be more obvious growths of lichens and small cushions of moss. In the maritime Antarctic, where conditions are less harsh, the mosses form extensive mounds and banks and even grass is found growing. The mosses of the continental Antarctic are much more sparse and grow no bigger than the size of a drinks coaster. Living in among the growths of mosses, algae and microbes are microscopic invertebrates such as nematodes, tardigrades and rotifers (Figure 2.6). Microarthropods – mites and springtails – browse on the growth in surprising numbers. These simple communities of organisms rely on local conditions to provide the opportunity for their survival and growth. Even among the largely barren soils of the Dry Valleys, there are microhabitats which provide conditions favourable for life. Although the air temperature may be freezing, a dark rock absorbs enough heat from the sun for any ice beneath it to melt and for liquid water to become available.

The vegetation of the continental Antarctic consists of mosses, lichens and algae, which develop in the moister areas. Lichens are more widespread than mosses. Crustose lichens cover the surface of rocks and pebbles like smears of paint. Foliose (leaflike) and fruticose (shrublike) lichens may have growths which project above the surface of the soil. Mosses and lichens have been recorded as close to the pole as 84°S

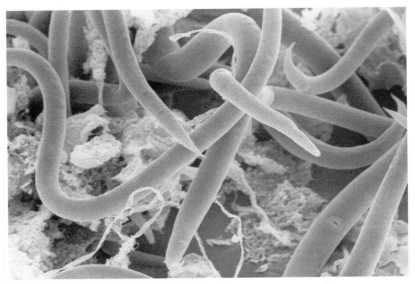

FIGURE 2.6 Scanning electron micrograph of an Antarctic nematode (*Panagrolaimus davidi*). The worms are about 1 millimetre long.

and 86° S, respectively. Growths of algae are relatively abundant in coastal areas where there is a seasonal supply of water from melting snow, and in and around shallow melt streams and ponds. In drier areas, vegetation can still occur under small rocks and pebbles which are transparent enough to allow sufficient light for photosynthesis. Stands of vegetation are more extensive in the milder conditions of the maritime Antarctic. In favourable places, mosses form turfs, banks and carpets and there are even two species of flowering plants: the Antarctic Hair Grass (*Deschampsia antarctica*) and the Antarctic Pearlwort (*Colobanthus quietensis*).

Microorganisms, such as algae, fungi, yeasts, bacteria and actinomycetes, occur in association with the plants but also in areas which do not support visible vegetation. The microbial communities are dominated by microscopic algae and cyanobacteria which are the most predominant photosynthetic organisms. These provide food for browsing springtails and mites in areas where there is no obvious vegetation. Microbes also colonise other specialised Antarctic habitats. In very dry areas, microbes can still survive by living within fissures or cracks in rocks (chasmoendolithic) or actually within the rocks, beneath its surface (cryptoendolithic), finding protection against desiccation and other hazards of their environment. These endolithic communities consist of a variety of organisms including algae and fungi (which are often closely associated and may form lichens) and cyanobacteria. These organisms lie dormant in a state of cryptobiosis for most of the time and can only grow when the occasional snow falls and its melt supplies them with sufficient water.

Some microorganisms even colonise habitats in snow and ice. Snow algae, which grow on the surface of snow and ice, are well known from more temperate areas of the world but are rare in the Antarctic, with most reports coming from the maritime Antarctic. Here, they produce patches of red-, green- or yellow-coloured ice and snow due to the presence of millions of algal cells, threads or spores. Wind-blown dust and debris can accumulate in holes on the surface of glaciers. As this material is dark, it absorbs heat from the sun and produces melting. This deepens the hole, allowing it to collect more debris and may expand to form pools several metres across. The hole also collects meltwater running over the surface of the glacier. These holes and pools are colonised by algae and cyanobacteria. Systems of melt ponds and streams also form on the surface of ice shelves and sheets during summer. These are colonised by algae and cyanobacteria and even by microscopic invertebrates such as nematodes and rotifers. Microorganisms have even been isolated from the surface of the central Antarctic ice plateau. These are mainly inactive forms which have been blown there, but it is possible some microbes could have brief periods of activity during the year. Living bacteria have even been isolated at the South Pole.

Summer meltwater from snow and glaciers forms streams on ice-free land. In some areas, the streams unite and provide a sufficient flow of water to justify calling it a river. The best known of these is the Onyx River in the Wright Valley of the Dry Valleys, which flows for up to three months a year in summer. The Onyx River is fed by glacial meltwater and flows away from the sea, discharging into Lake Vanda.

Antarctic lakes, and also streams and rivers which receive regular flows, provide perhaps the most favourable habitats for life in the continental Antarctic. Lake Vanda is 75 metres at its deepest point and is permanently covered by a sheet of ice up to 4 metres thick. Despite air temperatures which do not rise above 5 °C, and which remain below freezing for most of the year, the temperature of the lake increases with depth, reaching 25 °C at the bottom. The lake acts like a giant solar heater, absorbing heat from the sun during summer and trapping it beneath its layer of ice during winter. As well as changing in temperature with depth, there are also changes in salinity. The freshwater flowing in from the Onyx River lies on top of the saline water at the bottom of the lake.

Both freshwater and saline lakes and ponds are found in Antarctica. Some saline lakes were formed from seawater trapped as the ocean receded, others from the concentration of dissolved salts by the evaporation of water. The lakes of the continental Antarctic are permanently covered with ice, although the ice may melt around the edge of the lake during summer. Lakes in the maritime Antarctic may be free of ice for several months each year. A variety of organisms inhabit Antarctic lakes both in the water column and at the bottom of the lake. Most vegetation occurs as thick mats of algae and cyanobacteria at the bottom with protozoa and microscopic invertebrates (rotifers, nematodes and tardigrades) living in these mats. Some lakes have aquatic crustaceans, such as fairy shrimps, which can feed in the water column.

Lakes of unfrozen water even occur under the central ice plateau. These were first discovered in the mid 1970s. The largest is Lake Vostok, which is roughly the size of Lake Ontario and is covered by 4 kilometres of ice. How Lake Vostok formed is unknown. It may have been sealed by the formation of the polar ice sheet or it may have formed due to the melting of ice by its movement over the bedrock. Estimates of the age of Lake Vostok vary from hundreds of thousands to millions of years and it may well have been isolated from the atmosphere for all that time. It is perhaps the most extreme, isolated and pristine environment on Earth and may contain unique forms of life.

Studies of ice cores from the ice above Lake Vostok have yielded viable microorganisms that are 200000 years old. Scientists are currently trying to develop drilling techniques which will enable them to collect samples from the lake without contaminating it.

Cold is the most obvious challenge to life in the Antarctic, but, as well as being cold, most of Antarctica is very dry. This may seem surprising given the amount of ice and snow associated with the continent. However, frozen water is not available for organisms to use, unless they can melt it, a feat which few organisms can spare the energy to achieve. It is the availability of liquid water that determines where life can exist. During the brief summer, temperatures can rise sufficiently to melt accumulations of snow and partly melt glaciers and the edges of frozen lakes and ponds. This supplies enough liquid water for life. The supply of water is, however, transient and terrestrial Antarctic organisms have to survive periods of desiccation. Conditions favourable for growth may only exist for a few weeks or even days each year. Desiccation may, in fact, protect against freezing, since in a desiccated organism there is no water to freeze. However, during the summer, liquid water is present which, when it freezes, may result in the freezing of the organism. Freezing, or the threat of freezing, and freeze/thaw cycles are thus major hazards for terrestrial Antarctic organisms. Other threats include wind, unstable substrates and high levels of solar radiation (especially ultraviolet radiation, made worse by the ozone hole over the Antarctic). Resistance adaptation predominates among terrestrial Antarctic organisms; they lie dormant during adverse conditions and become active when conditions are favourable. There are, however, some examples of capacity adaptation. Antarctic lichens, for example, can continue photosynthesis at temperatures as low as $-10\,^{\circ}$C.

Antarctica has an extremely sparse flora and fauna. Only 74 plants and other macroflora and 186 animal species have been recorded from the maritime Antarctic (by 1993) and much fewer from the continental Antarctic. Geographical isolation is thought to be the major cause of this low biodiversity. Antarctica has had no land bridge with any other continent for more than 25 million years. The nearest land mass, South America, is over 1000 kilometres away. As well as distance from other land masses, the Southern Ocean and the circulation of air around the continent act as barriers to potential immigrants. The only route for immigration (other than largely accidental human introductions) is for seeds and other propagules to be transported by air currents or by birds. Any immigrants that do arrive have to establish themselves in the face of the harsh environmental conditions.

The study of life in Antarctica is important for a number of reasons. The communities of organisms consist of only a few species, making it feasible to attempt an overall understanding of their ecology. The organisms are adapted to survive some of the most extreme environmental stresses on Earth, making them good models to study how life can exist in extreme conditions and perhaps providing us with new techniques for storing and preserving biological materials. Since they are living at the limits of life on Earth, any change in their growth patterns may be a particularly sensitive indicator of global climate change. Polar regions may also warm first and so any effects of climate change could be seen earlier here than in other regions of the Earth. Finally, some Antarctic habitats are among the closest analogues we have on Earth to other bodies in our solar system which may contain life. The Antarctic has proved an important location to develop techniques which could allow us to search for and recognise the presence of life on other planets.

Antarctic marine organisms

The marine environments of the Antarctic are less extreme than their terrestrial counterparts. Marine organisms, by and large, are not exposed to desiccation and the thermal buffering effect of the water restricts variations in temperature. The marine environment is, however, not without its hazards. Organisms have to cope with low temperatures, the presence of ice, low light levels during winter, seasonal food supply and, in some places, high salinity.

The melting point of seawater is $-1.9\,°C$ and organisms which live close to sea ice have to function at this temperature. The presence of ice is ubiquitous in many marine Antarctic habitats, both as solid bodies of ice and as crystals floating in the water. The body fluids of marine invertebrates have much the same composition as seawater. They would freeze at the same temperature as seawater and are thus not at risk of freezing, unless the sea froze solid. Most Antarctic fish, however, have body fluids which are more dilute than seawater. This, together with the low temperature and presence of ice crystals, means that they are at constant risk of freezing. How they cope with this hazard will be explored in Chapter 5. Low temperatures slow the rate of biological processes and many marine organisms are slow growing and long lived.

Seasonal variations in both temperature and light intensity are major features, particularly for marine organisms living close to the continent. The temperature affects the extent of the pack ice which in winter covers an area seven times that in summer, forming a belt up to 1900 kilometres wide completely encircling the Antarctic continent. The pack ice not only affects the temperature of the water immediately beneath it but also the penetration of light, particularly if it is covered by snow. In winter, there is permanent darkness for several weeks. Photosynthesising organisms, such as algae, thus have to survive over winter with little or no light. When ice forms, the salt remaining becomes concentrated in unfrozen pockets, exposing any organisms trapped within them to very high salinities.

The seashore of Antarctica is relatively barren since much of it is permanently covered by ice or is kept clear of life by the scouring action of floating ice crashing against it. In some sheltered areas, the shore may be colonised by ephemeral seaweeds and by mobile molluscs which can escape ice damage by migrating. Permanently attached organisms, such as mussels and barnacles, which are common on seashores in other parts of the world, are not found on Antarctic shores. The actions of ice on the land produce large quantities of sediments which are deposited into the ocean. A rich community of organisms develops on these sediments where they form stable areas of the seafloor (benthic organisms). Conditions for benthic organisms are fairly stable, with low but steady temperatures and constant salinity. These organisms rely ultimately on phytoplankton from the waters above them as a source of food. This input of food is large but short lived, generally limited to when there is open water in summer. In contrast to the land, the ocean floor supports a rich and diverse community, especially in relatively shallow coastal waters.

The open waters of the sea are very productive, supporting a relatively simple food chain. Phytoplankton, mainly diatoms and algae, are fed on primarily by krill – a shrimp-like crustacean. Krill is a key organism in the food chain, being eaten by whales, seals, fish, squid and birds. Trapped within, or in close association with, the sea ice are diatoms, algae, protozoa and bacteria. The irregular surface under the ice can support thick growths of diatoms and other microorganisms. These live at the water/ice interface, become packed into channels within the ice itself or hang beneath it. Sea ice communities are thought to contribute substantially to the productivity of the phytoplankton by extending the period of the year over which it is active and providing a population which is liberated into the open water when the ice melts. The marginal-ice zone, at the edge of the receding pack ice, is thought to be particularly significant in this respect.

Most marine mammals (seals) and birds (penguins, skuas and other seabirds) that breed on the Antarctic continent do so only in summer and avoid the winter by migrating north. A dramatic exception is the Emperor penguin. This is the largest of the penguins and its large size helps it cope with the cold. Its size, however, poses a problem in rearing its chicks. Penguin chicks cannot go to sea and feed themselves until they are fully grown. The summer is too short for Emperor penguin chicks to complete their development, as the chicks of small penguins do. Emperor penguins have solved this by breeding before the start of the winter. The birds spend the summer feeding at sea, building up large stores of fat. The single egg is laid in May to early June, at the start of the winter. The females return to the sea shortly after laying the egg which is transferred to the male bird to incubate. The male holds the egg on its feet and covers it with a fold of skin and feathers which pro-

FIGURE 2.7 Male Emperor penguins incubating eggs. At over a metre tall, this is the world's largest penguin. Drawing by Jo Ogier.

tects it against the weather. And there the males stand, on the surface of the sea ice, incubating the egg during the depths of the winter and awaiting the return of the females some two months later (Figure 2.7). The chick hatches around the time the females return and are ready to go to sea by January or February when much of the sea ice has melted and food is readily available.

How do the males survive their long winter vigil? They do not feed, since they cannot abandon their egg to reach the distant open sea. They rely on their food reserves built up during the summer. Standing on the sea ice, rather than the land, probably helps since the sea beneath the ice (which is at $-1.9\,°C$) keeps them considerably warmer than they would be on land. The penguins are superbly insulated by their feathers and by the layer of fat beneath their skin. Their large size also helps them to retain heat. The males form creches and huddle tightly together, decreasing their exposed surface by as much as five-sixths and further reducing heat loss. They generate heat internally by utilising their food reserves. Their heat retention mechanisms are so efficient, however, that they only lose 15 per cent of their body weight during the nine weeks of incubation.

Fur, feathers and fat beneath the skin also provide the insulation needed by other penguins, seabirds and seals during their summer breeding season on land and during their activities in the water. Penguins are so efficiently insulated that on a sunny day they are in serious risk of overheating. They deal with this by increasing heat loss from their bodies by fluffing up their feathers and extending their flippers. Most birds do not have feathers on their legs and feet and a bird standing on cold ground or ice is at risk of losing heat through their feet. Indeed, if their feet were as warm as the rest of their body, they would melt the ice beneath them and become trapped as it refroze. The legs and feet are allowed, therefore, to cool to a lower temperature than that of the rest of the body. This is partly achieved by a countercurrent mechanism in which cold blood in the veins returning from the legs picks up heat from the warm arterial blood. The veins and arteries in the legs are close together to give the maximum area for this heat transfer to occur and thus to conserve heat within the body.

THE ARCTIC

The problems facing organisms in the Arctic are similar to those of Antarctic organisms. The main features of their environment are cold and the risk of freezing, a short growing season and a restricted supply of free water. The Arctic, however, supports a much richer and more diverse community of organisms than does the Antarctic. The primary reason for this is that the Arctic is an ocean surrounded by land (Figure 2.8), while the Antarctic is land surrounded by ocean. This has a number of important effects on the organisms that live there. The Antarctic is isolated from other land masses by the Southern Ocean, whereas the land bordering the Arctic ocean is, of course, continuous with land from more temperate regions. Land animals thus have more opportunity to migrate and to escape extreme conditions in the winter. The Arctic was also more easily colonised than was the Antarctic. As the glaciers retreated after the last Ice Age, organisms could invade the Arctic from more southerly regions. This process of colonisation is probably far from complete and, if interglacial conditions persist, the

FIGURE 2.8 Map of the Arctic region, showing the Arctic Circle and the approximate position of the treeline (the northern extent of coniferous forests) and areas of tundra.

Key: 🌲 treeline   ⁖ tundra

diversity of the Arctic biota will become even greater than it is at present. Climatic conditions in the Arctic are less extreme than those in equivalent latitudes in the Antarctic. This is because the Arctic ocean retains heat better than the land mass of the Antarctic and warmer air and water are carried to the Arctic from more temperate regions.

The Arctic is generally defined as the area north of the treeline, where the great coniferous forests of Europe, Asia and North America cease. This corresponds to the area where average summer monthly temperatures do not exceed 10 °C. The dominant environment is tundra – large tracts of treeless open land north of the Arctic circle where the ground thaws during summer. Tundra is a form of grassland and covers one-tenth of the Earth's land surface. The main types of vegetation are grasses, sedges, reeds and woody shrubs such as heather. There are over 400 species of flowering plants (the Antarctic Region has only two). Mosses and lichens are common and form a ground cover beneath the other vegetation. Although the Arctic is defined as being north of the tree line, there are in fact two species of trees – the dwarf birch and the Arctic willow – which form low-growing shrubs.

All Arctic plants are dwarfed and grow close to the ground. This protects them from the fierce wind and from blasting by snow and dust. It also means they are more likely to be covered and protected by snow during winter. Temperatures close to the ground are generally higher than air temperatures. Other adaptations which tend to raise the temperature of plants include the dark colouration of leaves and flowers, the dense growth of stems and leaves, hairs which trap an insulating layer of air and the parabolic shape of some flowers which tends to concentrate the heat of the sun. The roots of the plants may be restricted to the upper layers of the soil which are warmer and better drained.

The vegetation can only grow when the sun melts the snow and the surface of the soil in summer. The water is prevented from penetrating into the deeper layers by the permafrost (permanently frozen ground) beneath the soil surface. In spring, the soil surface becomes saturated with water, providing for a flush of growth. If it were not for the permafrost, the spring meltwater would rapidly drain away and plants would not be able to grow. The soil dries out in summer, however, and the vegetation then has to rely on rainfall. This is generally low, resulting in semi-desert conditions in many places. The snow is not only important for providing water in the spring but it also protects the plants and animals beneath it from the harsh winter conditions. However, plants cannot grow much until the snow cover melts. The growing season is thus short and plants have to flower and set seed very rapidly. They are helped in this task by the continuous daylight that prevails for one to four months over summer. The flush of spring growth provides abundant food for animals.

There are over 2000 species of free-living arthropods in the Arctic (including spiders, mites, springtails and insects). This compares with about 140 in the Antarctic (mites, springtails and two midges) and over 25 000 in the United Kingdom. The largest group of insects are the Diptera (flies), with over 50 per cent of insects recorded from Arctic North America belonging to this group. Like the plants, Arctic insects only have a short period of the year in which they can grow. Many different life cycle patterns are found, with insects overwintering as adults, eggs, larvae and pupae. Most take two or more years to complete their life cycles. An extreme example is a moth, *Gynaephora groenlandica*, which lives in Arctic areas of Canada and Greenland. At Ellesmere Island in the Canadian Arctic (which is 78° N), their life cycle can take up to 14 years to complete. There is no special stage for surviving the winter and the moth can overwinter as any one of a number of different larval stages. The larvae survive the winter in a frozen state and resume their development for a brief period during the summer. Other insects, however, can complete their development in one season. At Lake Hazen, also on Ellesmere Island, mosquitoes lay their eggs in south-facing sites. These sites clear of snow first in the summer and the eggs receive the maximum amount of sun, allowing their rapid development and hatching.

Arctic insects tend to be small, dark and hairy – or at least more so than their relations from more southerly regions. This helps them gain and retain heat. Insects bask in sunny sites and on flowers which concentrate the heat of the sun. Blowflies lay their eggs on carcasses that are exposed to the sun and the development of maggots is always faster on the sunny side of the carcass. Arctic bumblebees are the only Arctic insects that can produce their own heat – by rapidly vibrating their wing muscles. The metabolism of Arctic insects is adapted to the conditions and they can develop and grow at lower temperatures than their temperate relations. The wings of Arctic insects tend to be reduced or are absent altogether. This may help them to stay put in windy conditions or perhaps they do not have enough time or resources to devote to the development of wings during their brief growing season. They have various mechanisms for surviving subzero temperatures (see Chapter 5). Other Arctic land invertebrates include nematodes, rotifers, tardigrades and annelids.

Phytoplankton, including diatoms associated with the ice, are the primary producers of the Arctic ocean. These are consumed by zooplankton (especially crustaceans such as krill) and the food resources of the ocean are exploited by seals, walruses, whales, polar bears and a variety of seabirds. Members of the auk family, guillemots, razor-bills, puffins and auks themselves have a similar lifestyle to the penguins of Antarctica, although they have retained their ability to fly. Open ice-free areas of water (polynas) are important for marine mammals and birds, particularly polynas which are recurrent and remain open during the winter due to local conditions of wind and currents. Polar bears feed mainly at sea (on seals), but in the summer they also feed on land, catching lemmings and nibbling berries.

Unlike the Antarctic, the Arctic supports several terrestrial mammals and birds. The majority of birds migrate south to avoid the harsh winter but return to the Arctic to take advantage of the spring growth of plants and the emergence of vast swarms of insects. Permanent bird inhabitants include the ptarmigan and snowy owls. Both these birds are well insulated by feathers; indeed, snowy owls have the best insulation of any bird. Ptarmigan burrow into the snow at night to reduce their heat loss.

Only a few Arctic mammals, such as the Alaska marmot and Arctic ground squirrel, hibernate to avoid the winter. Arctic hares, lemmings and other small rodents like shrews and voles remain active throughout the winter, surviving by burrowing into the snow. Their small size means that they cannot themselves control heat loss sufficiently to survive the winter, but snow is an excellent insulator and temperatures beneath the snow are maintained at -10 to $0\,°C$, despite much lower air temperatures. Light may even penetrate beneath the snow, allowing grass to remain green and provide food for the animals. Large herbivores such as caribou, musk oxen and reindeer rely on vegetation on ground which is free of snow, or where the snow cover is sufficiently shallow for them to expose the vegetation by digging. Musk ox are the large mammals perhaps best adapted to the harsh conditions of the Arctic tundra and polar desert; it is the only one which is restricted to these regions. Its adaptations for conserving heat include a large size, low compact body shape and a dark, thick, woolly coat, which hangs beneath its body forming a protective curtain. Herds of caribou migrate south to spend their winter in the forests, accompanied by the predators which prey on them.

Mammalian predators include the Arctic fox, wolves, wolverine, ermine and stoats. Arctic foxes are so well insulated that they can sleep on open snow at temperatures down to $-80\,^{\circ}\text{C}$ for up to an hour without their core body temperature decreasing. They can generate heat internally, but the rate of metabolic heat production is not increased until the ambient temperature falls below $-40\,^{\circ}\text{C}$. Both predators and prey grow white coats to camouflage themselves against the winter snow.

MOUNTAINS

Mountains are the only other environments, apart from polar regions, where organisms are exposed to permanent snow and ice. Temperature decreases with increasing altitude due to the thinning of the atmosphere. The air is warmed by the transfer of energy from the sun. As it becomes thinner with increasing altitude, there is less air to absorb this energy and hence it is colder. The temperature decreases by an average of $6.5\,^{\circ}\text{C}$ for every 1000 metre increase in altitude. The decrease in temperature results in permanent snow cover even on mountains at the equator, such as Mount Kenya in East Africa and Cotopaxi in the Ecuadorian Andes. You can even ski on the snowfields of Mauna Kea in the Hawaiian Islands.

By forcing moving air to ascend, mountains cause it to lose its moisture in the form of rain and other forms of precipitation. Most of this falls on the lower slopes and the air at high altitude is thus very dry. The low moisture content of the air, and the lack of dust and other particles, means the air is very clear. This allows the passage of high levels of solar radiation, particularly ultraviolet radiation. The thin air does not absorb much of this radiation, but the ground does, resulting in marked differences in air and ground temperatures. The ground loses its heat rapidly at night and there are thus wide variations in temperature between day and night. Since there is less contact between the air and the ground, winds reach high speeds on mountain tops. This contributes to high rates of evaporation and to low temperatures. The thin atmosphere at high altitudes also poses oxygen-supply problems for organisms.

The severity of the mountain environment increases with increasing altitude and several distinct zones can be recognised. Mountains are found throughout the world, however, and conditions will vary according to how close they are to the equator or poles and to oceans. In temperate regions, the lower slopes may be covered in deciduous forest, followed by a zone of coniferous forest. The treeline, above which trees do not grow, corresponds to where average temperatures in the warmest month of the year are below 10 °C – the same conditions which determine the location of the treeline bordering the Arctic. Above the treeline is alpine grassland or tundra which is bordered at a higher altitude by the snowline.

The tops of high mountains are so cold and windswept that there is little life associated with them. Apart from the occasional climber, the only life to be found may be insects, birds and microorganisms blown there from lower altitudes. Lower down, the snow may be tinged green or pink by the presence of snow algae. These have a mobile stage in their life cycle which enables them to migrate through the snow to a level where they receive enough light for photosynthesis, but remain below the surface where they are protected against the extreme conditions. In the lowland forests of Central Europe, the surface of the snow is sometimes stained black by masses of springtails. Other insects and even earthworms and enchytraeids are found in among the snow.

Bacteria have even been found beneath glaciers in the Swiss Alps, where their activity is responsible for eroding rocks. Productivity at high altitude is low and many of the organisms rely on material which is carried there from the lowlands. This material consists of organic matter, such as seeds, leaves, pollen grains, spiders and insects, usually dead and known collectively as aeolian (windborne) derelicts.

The vegetation of alpine grasslands or tundra superficially resembles that found in the Arctic tundra. This is not surprising since they face some similar environmental challenges – strong winds, low temperatures, a short growing period and low moisture levels. The two environments are, however, different and alpine vegetation faces the additional hazards of high ultraviolet radiation and, in tropical areas particularly, wide daily fluctuations in temperature with freezing conditions during the night but heat during the day. Seasonal changes are small on tropical mountains, but, at high altitude, summer comes every day and winter every night. In the high valleys on the slopes of Mount Kenya, this has led to some peculiar adaptations (Figure 2.9). Giant lobelias and groundsels grow there, the latter looking like giant cabbages on trunks. A covering of hairs or dead leaves on these plants traps a layer of air which insulates the plant against temperature extremes. Giant plants, the frailejones and puya, are also found in the Andes of Bolivia and Peru. It is not known how their large size helps these plants to survive, but it must be related to the proximity of both these sites to the equator. Perhaps favourable growing conditions during the day at the equator enables them to reach a large size that helps them survive the cold during the night. Many alpine plants, however, are small and grow close to the ground. They have extensive root systems to anchor them against the wind. The leaves often form rosettes so that they do not shade each other from the sun. The production of a downy or woolly covering protects the plant against ultraviolet radiation (by reflecting it), retards water loss and provides thermal insulation. Protection is also provided by dried leaf bases and scales. The vegetation is often blue–green or yellow–green (rather than green) in colour which reflects harmful infra-red and ultraviolet radiation.

FIGURE 2.9 Giant lobelias and groundsels growing on Mt. Kenya. These plants grow to over 6 metres tall. Drawing by Jo Ogier.

Alpine insects have many of the features observed in Arctic species. They are small and dark, can survive low temperatures and have a tendency to become wingless and to have protracted life cycles. Winglessness has developed for similar reasons to Arctic insects, but also because alpine insects are in particular danger of being blown away and deposited in an unsuitable habitat. Dark colouration may help protect the insect against the intense ultraviolet radiation and enable it to absorb heat to raise its temperature to a level where it can maintain activity. Insects may become encased in ice at high altitude, which can restrict their oxygen supply. A number of alpine species can survive periods of anoxia and can function without oxygen. Spiders and mites are the other main terrestrial invertebrates. Other invertebrates (flatworms, nematodes, rotifers, molluscs and crustacea) are particularly associated with alpine lakes and streams.

The small size of invertebrates helps them gain sufficient oxygen at high altitude despite the thin air. Mountaineers usually carry oxygen with them at altitudes above 6000 metres. Alpine mammals, however, have to ensure a sufficient supply of oxygen to their tissues by increasing the ability of their blood to carry oxygen. Vicuñas, relatives of the camel, have large numbers of small oxygen-carrying red blood cells. The blood of vicuñas has 13 million red blood cells per millilitre, compared with 5 million in humans, and their blood can carry about 25 per cent more oxygen than human blood. Alpine mammals are well insulated against the cold. Chinchillas, rodents from the Andes, produce the most dense and silky fur of any animal in the world and the coat of vicuñas is the most highly prized of wools. The large size of some animals from high altitudes, such as the yak, also helps them conserve heat.

THE BLEAK MIDWINTER

As we descend the mountains or move south from the Arctic tundra, the conditions experienced by organisms through the year become less extreme overall and we encounter temperate conditions. The most noticeable change is that this allows the growth of large trees. Trees need at least 30 days in the year when temperatures are at 10 °C or higher and, when there is sufficient light, in order to develop and construct their trunks. Conditions at other times of the year, however, can be just as harsh as those experienced on the Arctic tundra, with temperatures falling below −40 °C and the ground covered with a thick blanket of snow. This not only threatens to freeze organisms' cells but slows growth and denies organisms access to liquid water. Organisms which live in temperate regions must survive these bleak winter conditions.

The forests just below or south of the treeline consist solely or largely of coniferous (cone-bearing) trees. The leaves of conifers are reduced to thin needles. Snow does not easily settle on such leaves and this helps prevent branches being broken by the weight of snow. The spines contain only a little sap, which is sugary, making freezing less likely. The tree is protected against water loss by the size, shape and thick waxy covering of their needles. Pine needles have relatively few stomata and these are protected at the bottom of pits which lie in grooves along the length of the needle, reducing water loss. When the ground is frozen, cutting off water to the tree, the stomata close.

Most conifers are evergreen, retaining their needles throughout the year. The larch, however, grows in areas which are dry, as well as cold in winter, and cannot afford to lose any water during winter. It sheds its leaves and becomes dormant. This is a much more efficient and reliable way for a tree to survive the winter. Where the summers are long enough for them to grow and set seed in one season, the forest is dominated by deciduous broad-leafed trees. Their leaves are much more efficient at gathering light than are those of conifers, but they are too susceptible to water loss, freezing and wind damage during winter to remain on the tree. As winter approaches, nutrients and chlorophyll are withdrawn from the leaves, this process being triggered by the shortening days and falling temperatures of autumn. This reveals the waste products of photosynthesis that remain and the leaves turn red, yellow or brown. A blockage forms at the base of the leaf stalk, sealing it off until it withers and detaches from the tree. The tree then sits out the winter in a dormant state. Shedding and regrowing leaves each year is demanding of resources and deciduous trees can only grow where there is a sufficient supply of nutrients and a reasonably long summer.

Many animals also survive the winter in a dormant state. A few insects remain active throughout the winter but become inactive (quiescent) during periods of particularly low temperatures. Some show longer periods of dormancy. Perhaps the most sophisticated response to winter conditions is to enter a period of diapause. The insect anticipates the onset of winter by entering a long-term dormant state (diapause) in which it remains over winter before resuming activity in the spring. This usually only occurs in one stage of the life cycle, but, for different species, this may be the egg, larva, pupa or adult. The key feature of diapause is that it is not triggered by the harsh conditions themselves but by changes which indicate that the harsh conditions are approaching. One of the most important cues that trigger the entry of the insect into a state of diapause is photoperiod (the relative length of day and night). During the autumn, the days get shorter and the nights get longer. This change in photoperiod is a more reliable indicator of the onset of winter than changes in temperature, which may be only temporary. The change in photoperiod acts directly on the brain of the insect, triggering the production of hormones which control the changes that result in diapause. Both diapausing and non-diapausing insects have mechanisms which allow them to survive subzero temperatures. These will be explored in Chapter 5.

Fish and amphibians have little ability to regulate their internal temperature while reptiles can only do so by behavioural mechanisms, such as basking in the sun. These ectotherms are generally at the same temperature as their environment and, when temperatures fall, they lose heat and become inactive. The survival of most species over winter depends on choosing overwintering sites where they are protected against freezing, such as deep within the ground and at the bottom of deep lakes and ponds. Some can survive freezing (see Chapter 5).

Many birds and some large mammals migrate to warmer climates to avoid a harsh winter. Small mammals, however, must remain and survive. They cannot conserve their body heat as well as a large mammal can, given their larger surface area in relation to their volume. Their ability to develop a thick coat as insulation is also limited by their size, since a thick coat would limit the movement of a small animal. If the fur of a mouse was as thick as that of a musk ox, its feet would not reach the ground. Small mammals can compensate, however, for their limited insulation by burrowing beneath the snow and by forming lairs, burrows, dens and other forms of shelter which insulate them from the external environment. Most mammals maintain their body temperature at about 37–38 °C. They do this by generating heat metabolically and by measures for conserving heat. To generate heat, the animal needs to consume food and many mammals remain active throughout the winter if they can find sufficient food to do so. They will, however, undergo periods of dormancy of varying degrees and duration. The function of dormancy is related to the need to conserve food during periods when it is scarce, rather than the survival of low temperatures *per se*.

Many birds and mammals, including ourselves, undergo periods of dormancy in which they become inactive and there is a drop in body temperature. Birds and mammals active during the day sleep at night. In humans, oxygen consumption drops by about 10 per cent and body temperature by 1–2 °C. Even these small changes result in considerable savings in energy for an endothermic animal, of about 7–15 per cent, which conserves their food resources. Animals are easily aroused from sleep by disturbance, but some enter deeper periods of dormancy. During torpor, a more profound dormancy, body temperature drops lower than it does in sleep (to between 10 °C and 30 °C). The animal has to restore its body temperature before it can become active and thus takes longer to arouse (a few hours). Torpor may occur on a daily basis, in humming birds for example, or for longer periods.

Hibernation (winter sleep) is certainly a way of avoiding harsh winter conditions, but it is much less widespread than most people think. Grey and red squirrels are active throughout the winter and maintain a high body temperature. They rely on stored food reserves and spend most of their time in their nests and are thus seen less during winter. Few Arctic mammals hibernate and lemmings and Arctic rodents remain active beneath the snow. Hibernation requires the build-up of food reserves and Arctic mammals cannot eat enough food during summer to achieve this. Some bears (such as brown, grizzly and Himalayan bears) spend the winter in dens. These may be natural caves, holes dug out of hills or beneath the roots of large trees. In winter, they often become covered by snow, which improves their insulation. Female polar bears dig large snow caves, but only when they are to produce cubs. Brown bears may spend 3–5 months in their dens, even giving birth to cubs before emerging in spring. The bears are, however, easily aroused from their dormancy within the den, as some investigators have found to their cost, and it is a form of sleep or torpor rather than a deep hibernation.

Deep hibernation (sometimes called true or classic hibernation) involves a decline in temperature to within a few degrees of the ambient temperature. Body temperatures as low as $-2.9\,°C$ have been recorded from hibernating ground squirrels. A wide range of small mammals (and a few birds) exhibit deep hibernation. They prepare for hibernation by eating in excess of their daily requirements and storing food in their bodies in the form of fat. This fat is the main source of energy while they are hibernating, although some species store seeds and nuts to eat during periods of arousal. During entry into hibernation, the animal retreats to its burrow and its body temperature falls over a period of several days. The hibernating site is usually sealed and this promotes a build-up of carbon dioxide which helps depress the metabolism of the animal. Hibernating animals have a low level of metabolism, low temperatures, the heart rate falls and they do not drink, defecate or urinate. Hibernation results in considerable energy savings. The energy consumption of hibernating hedgehogs (at $5.2\,°C$) is 96 per cent below that when they are active. Some hibernating animals have periods of arousal, when they raise their body temperature and resume activity for a brief period. For those which store food supplies, this enables them to feed, but others do not store food and the reason for arousals in these species is something of a mystery.

IN THE DEPTHS
The cold deep sea
For those of us who rarely venture on the surface of the waves, it is difficult to comprehend how big the oceans of the world are. They cover nearly three times as much area of the Earth as does the land. The

oceans are a three-dimensional habitat. Their average depth is more than 3 kilometres and the deepest parts, the great ocean trenches, reach a depth of nearly 11 kilometres below the surface. This is much deeper than the largest mountains on the land are tall. Life is found pretty much everywhere in the oceans. Among other things, marine habitats vary with latitude, proximity to land, availability of solid substrates such as rock and depth. There is a great range and variety of ocean habitats.

Humans cannot survive beneath the surface of the sea for long without special equipment. The ocean depths are so outside our experience that it is difficult to judge what parts of its variety of habitats are extreme or not extreme. The availability of water is clearly not a problem for marine organisms and there is sufficient oxygen dissolved in most waters to supply their needs. Temperature is by and large no problem either. The enormous mass of water provides a great deal of thermal buffering. The surface waters are the most variable, reaching 40 °C in the shallow seas of the Arabian Gulf and the Red Sea. The lowest surface temperature is −1.9 °C in polar waters, determined by the freezing point of seawater. In the tropics and at the poles, there is little change in surface temperature with season. Temperate waters are more variable, with temperatures around the United Kingdom reaching 19 °C in September and falling to 2 °C in winter. Temperature declines with depth and deep waters below 2000 metres are at a constant 2–4 °C throughout the world.

Light is absorbed by seawater and, even in clear waters, only 1 per cent of the sunlight falling on the surface penetrates to a depth of 50 metres. Photosynthetic organisms are not found below about 100 metres. Most organisms living deeper than this have to rely on food in the form of detritus (dead organisms and other organic debris) drifting down from the surface waters. Pressure increases rapidly with depth and organisms living at the bottom of the oceans have to cope with enormous pressures. How they do so will be covered in Chapter 6.

Until recently, our knowledge of organisms living in the ocean depths has been extremely limited. Scientists trying to sample organisms from the depths have had to solve the technical difficulties posed by the distance from the surface and the high pressures involved. In the late 1960s, Bob Hessler and Howard Sanders from the Woods Hole Oceanographic Institute began investigating deep ocean beds using a sledge which was towed behind their ship, collecting samples into a net. They found that, rather than the muds of the deep ocean being a virtual desert, as had been previously assumed, they contained large numbers of animals. In the mid1980s, Fred Grassle, of Woods Hole, and Nancy Maciolek headed a team of scientists who spent two years collecting a series of cores from a depth of around 2100 metres off the east coast of the USA. Of the animals they found, 58 per cent were new to science. Each extra kilometre they sampled yielded a new species. Extrapolating from this, they estimated that the ocean depths could contain about 10 million species. This estimate is, however, based on sampling a very small proportion of the ocean bed and some think it is way too high. Others, however, point out that the estimates are based only on the numbers of relatively large invertebrate animals. Microscopic invertebrates, particularly nematodes, also occur in large numbers and most of those that have been recovered from the depths have proved to be new species. Estimates of the numbers of deep sea species thus vary from half a million to 100 million. To put this into context, about 160 000 marine species have been described so far and about 1.8 million for the whole Earth. It is clear that the deep oceans provide a substantial proportion of the Earth's biodiversity and that much of it remains to be discovered.

Scientists have also begun to explore the depths using submersible vehicles. Remote operated vehicles (ROVs) are engineered to withstand the high pressure and are controlled by, and send data via, a cable connected to a ship on the surface. In 1995, Japanese scientists succeeded in landing their ROV Kaiko at the bottom of the Mariana Trench in the Philippine Sea. This is the deepest ocean trench in the world and one which would easily accommodate Mount Everest if it were submersed upside down. Even here, the cameras attached to Kaiko soon observed life in the form of sea cucumbers and worms. The life visible on the surface of the mud is, however, little indication of what lies beneath. Samples brought back by Kaiko from the Mariana Trench contained millions of bacteria per gram of mud – much less than would be found in garden soil, but still a reasonably large number.

Both unmanned ROVs and manned submersibles are also changing our view of the great volume of ocean that lies between the ocean floor and the surface waters. Sampling this region with nets only catches a tiny proportion of the animals that live there. Many are too fragile to be transported from the water which supports them. Submersibles such as Deep Rover, a one-person submersible operated by the Monterey Bay Aquarium Research Institute, allow scientists to observe the life of this region directly. Most animals there are transparent and jelly like. Siphonophores belong to the same phylum as jellyfish, but make up colonies of animals which may be as much as 40 metres long. They act like living driftnets, catching particles of food in the water. Bruce Robinson, a scientist at the Monterey Bay Aquarium, describes the midwater as 'a dim, weightless world filled with ragged three-dimensional spiderwebs' and that the animals, and their discarded body parts, are so numerous that 'we now think of this delicate marine life as *forming* much of that midwater environment.'

Hot vents and cold seeps
In some places, life on the ocean floor does not have to rely on the detritus descending from the surface but can tap other, more unusual, sources of energy. In 1977, scientists from the Woods Hole Oceanographic Institute, exploring near the Galapagos Islands using their ROV Alvin, found the deep ocean vents which had been predicted by geologists. These are areas where material deep within the Earth comes to the surface, forming new seafloor and moving apart the great continental plates which make up the Earth's surface. Water penetrates these vents and emerges superheated, because of the high pressure, to temperatures as high as 350 °C. What was not predicted, however, was that these seemingly inhospitable hydrothermal vents teemed with life. They were surrounded by large numbers of giant tube

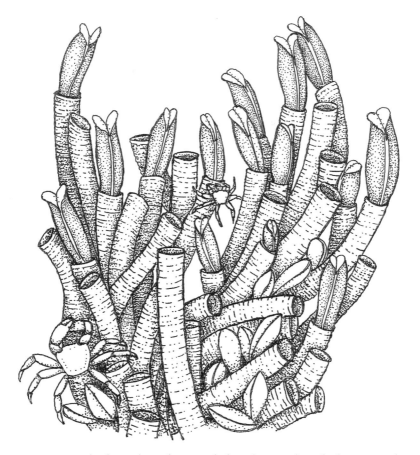

FIGURE 2.10 Hydrothermal vent fauna, including clams, crabs and tubeworms. The tubeworms grow up to 1.5 metres tall. Drawing by Jo Ogier.

worms, up to 11.5 metres long, giant clams, crabs and fish (Figure 2.10). More than 400 species of animals have been identified (by 1997), with molluscs, arthropods and annelids being the commonest groups. The tube worms were initially a bit of a mystery, being previously unknown to science. They turned out to be pogonophorans, a previously obscure phylum of animals only discovered in 1900. The pogonophorans around the vents are so different from others of the phyla that they are often given their own class or phylum, the Vestimeniferans. Vestimeniferans have no gut or intestinal system and no light penetrates to the depths where the hydrothermal vents are found. What do the tube worms and other animals feed on?

The hot water issuing from the vents is rich in minerals dissolved under high temperature and pressure from rocks beneath the surface. It often contains high concentrations of sulphides. Hydrothermal vents swarm with bacteria (archaea) which live by oxidising the sulphides (mainly hydrogen sulphide) to release their chemical energy. The animals feed on the bacteria or on other animals. The tube worms, and other animals, harbour the sulphur-oxidising bacteria within their tissues, feeding on the nutrients produced by their activities. This was the first community of organisms to be discovered which obtains its energy from chemicals (chemotrophic) rather than sunlight (photo-trophic). Although the water emerges from the vents at very high temperature (over 300 °C), it rapidly cools as it moves away to the temperature of the surrounding seawater (about 2 °C). The bacteria, however, need to be close to the vents to capture the sulphides before they become too dilute. They are extreme thermophiles and can tolerate temperatures as high as 113 °C. They also need access to oxygen (or to nitrous oxide) dissolved in the seawater surrounding the vents to oxidise the sulphides. The organisms inhabit a fairly narrow zone around the vents where their requirements for sulphides and oxygen can be met. The tube worms are bright red because of the haemoglobin in their blood; this transports not only oxygen (as does our blood), but also sulphides (which is unusual). The animals thus supply their bacterial partners with the chemicals they need. Some vent animals can tolerate temperatures up to about 50 °C, but most animals are associated with waters at temperatures below 30 °C. The hot waters from the vents and the cold waters of the surrounding ocean do not mix well, however, and the animals are exposed both to extremes of temperature and to rapid changes in temperature. They also have to tolerate the high concentrations of minerals, and other toxins, dissolved in the water.

In places, the minerals are deposited to form tall chimney-like structures. Black sulphide-rich water issues from the top of these, giving them the name 'black smokers'. Some organisms live on the walls of the black smokers, where they may be exposed to very high temperatures. The Pompeii worm (*Alvinella pompejana*), so called because it lives within the rain of volcanic material issuing from the black smokers, is a polychaete worm (an annelid, the same phylum as earthworms) that lives in tubes which it constructs on the outer walls of the chimneys. While the temperature of the water within the chimneys is very high, it rapidly cools when it meets the surrounding seawater. The Pompeii worm, however, lives very close to the scalding water and probes inserted into their tubes have measured temperatures as high as $81\,°C$ at the end closest to the chimney and $22\,°C$ at their opening. If these temperatures reflect those within the worm itself, this means that the Pompeii worm is the most thermotolerant animal known and one which is exposed to an extraordinary temperature gradient along its body of up to $60\,°C$.

In 1984, another strange deep-sea habitat was discovered. Erwin Suess of the Research Centre for Marine Geosciences in Kiel, Germany, also using the ROV Alvin, observed wedges of mud in the form of ridges accumulating where the Juan de Fuca tectonic plate slides beneath the North American plate off the coast of Oregon. These were dotted with stone chimneys formed of minerals from plumes of water and gas issuing from the seabed. However, unlike the hydrothermal vents, these waters were cold, giving them the name 'cold vents' or 'cold seeps'. If the temperature is low enough, the methane issuing from these seeps becomes trapped within an icy cage of water molecules, forming methane hydrates which look like lumps of dirty ice. Methane hydrates are also found on land in places in the Arctic. These hydrates have been shown to occur throughout the oceans of the world in enormous quantities and it is thought that the world's methane hydrate deposits contain twice as much carbon as all known coal, oil and natural gas deposits put together. They could be an important energy resource when other fossil fuel reserves are exhausted, if the problems of harvesting them can be overcome. These methane deposits are thought to have been formed by microorganisms decomposing the organic material in ocean sediments.

Melting of the methane hydrate releases not only methane and water but also hydrogen sulphide and ammonia. These chemicals provide an energy source for dense communities of chemotrophic bacteria, which, in turn, provide food for animals including clams and tube worms. The only animal which actually lives within the methane hydrate is the 'ice worm', *Hesiocaeca methanicola*, a species of polychaete worm. These worms create a current of water which gradually wears away the ice and form burrows in which they live, feeding on the bacteria. The methane ice is, however, an unstable place to live, since it melts above 6 °C and a small rise in temperature can cause it to disappear.

The underworld

Until about the late 1980s, most scientists believed that life was restricted to the top few metres of the soil or ocean sediments. As depth increased, nutrients became sparse and so did organisms. The few reports of organisms being recovered from great depths within the Earth were dismissed as contamination with material from the surface layers. Two technical developments changed this view. The first was the development of drilling techniques which gave confidence that cores could be retrieved from depth without contamination. Samples were recovered using a diamond-studded drill bit which headed a great length of rotating steel pipe from a drilling derrick. A concentrated tracer material was added to the lubricating fluid so that when the core of rock was removed any contaminated material could be identified and cut away to leave a pristine sample of rock from deep within the Earth. The second development was the advent of techniques for identifying microorganisms without having to grow them in culture. All organisms contain DNA and their presence can be revealed by dyes which either stain DNA directly or can be attached to nucleic acid probes. By varying the nucleic acid probe, the presence of different types of microorganism can be demonstrated.

The first scientists to use these techniques were involved in the Subsurface Science Programme of the US Department of Energy (DOE). They were interested in the possibility that, if organisms existed in the depths of the Earth, they might degrade organic pollutants and help maintain the purity of groundwater or, rather less usefully, degrade the containers in which the DOE was proposing to deposit the radioactive waste from nuclear facilities. They demonstrated the presence of many different types of microorganisms in rocks at depths down to 500 metres beneath the surface. Since then, microbes have been discovered in many different types of rocks and deep within ocean sediments. The record depth at which life has been found is at the bottom of a South African gold mine, 3.5 kilometres below ground. Pressure and temperature increase as you go deeper into the Earth. Bacteria from hydrothermal vents can grow at 110 °C and some scientists think that subsurface bacteria could withstand temperatures as high as 150 °C. This would allow organisms to exist to depths of about 7 kilometres beneath the seafloor and to 4 kilometres below the surface of the land. Although the organisms are often sparsely distributed, this is such an enormous volume that it has been estimated that the total biomass of deep subsurface organisms exceeds that of those living on, or just below, the surface.

Bacteria are the most numerous of these subsurface organisms, but there are also fungi and protozoa. Some 10 000 strains of microorganism have been isolated from subsurface cores. Each gram of rock contains anything from 100 bacteria to 10 million bacteria (compared with more than one billion per gram in agricultural soils); ocean sediments contain even higher numbers. The protozoa feed on the bacteria, forming part of a simple subterranean food chain, but what do the bacteria feed on? Sedimentary rocks are formed from sands and from ocean, river or lake sediments that have organic material trapped within them. Microbes living in pores within the sediments can utilise these ancient nutrients and grow. As sedimentary rocks are buried more deeply, they become increasingly compacted and their pores filled with minerals. The distribution of microorganisms is thus likely to become more patchy, condensed into the remaining pores and concentrations of nutrients. The bulk of the Earth's crust, however, consists of igneous rocks, such as granite and basalt, which are solidified from molten magma. These rocks were too hot to support life when they were first formed; the organisms which inhabit cracks and fissures within the rocks are carried there by the groundwater flowing through them. Subsurface bacteria do not just rely on nutrients trapped within the rock or carried there by groundwater. Some are chemotrophs, deriving their energy from the oxidation of iron or sulphur compounds and building organic material directly from the carbon dioxide and hydrogen gas dissolved in the rock. These bacteria excrete organic compounds which are then utilised by other types of bacteria. These ecosystems based on chemotrophic bacteria are completely independent of material and solar energy from the surface. They have been referred to by some scientists as 'SLiMES' (subsurface lithoautotrophic microbial ecosystems). Some of these communities of organisms have been isolated from the surface for a long time and are at least several million years old. Nutrients are in short supply in most parts of the deep subsurface and the organisms are likely to grow very slowly indeed, perhaps reproducing once every few hundred years. The bacteria are typically very small, reflecting the low availability of nutrients.

The discovery of life beneath the surface of the Earth has profound implications for our understanding of many of the processes which occur there. Changes in the deep subsurface, rather than being due to purely physical and chemical transformations, may also involve biological processes. Bacteria may be involved in the concentration of minerals, such as gold, into seams and methane-producing bacteria could be responsible for forming natural gas deposits. It has even been suggested by Tom Gold of Cornell University, in a controversial theory, that the world's oil deposits are formed by the activity of subsurface microorganisms, rather than from the remains of ancient plants and animals. This would mean that the formation of oil is a continuing process and that, rather than oil reserves being finite, they are being continually renewed. Many of the bacteria may break down organic material and be useful in the cleaning of contaminated soils and groundwater.

In places, material from the underworld makes an appearance at the surface. Bacterial flocks emerging from some deep sea vents originate from SLiMEs and, as we have seen, minerals emerge from hydrothermal vents and cold seeps. Material from the deep makes a dramatic appearance on land in the form of volcanoes. The violent eruptions of lava and ash destroy any life with which they come into contact, but, eventually, the eruptions cease, the volcanic material cools and it is colonised by organisms. Rather less violent are hot springs formed by groundwater or rainwater which has been heated through contact with lava and forced to the surface. The scaldingly hot water can nevertheless support the growth of extreme thermophilic bacteria and cyanobacteria (see Chapter 4). Volcanic activity can even make some otherwise inhospitable environments capable of supporting life. Mount Erebus on Ross Island in the Antarctic is an active volcano. Around the edge of the volcano, the temperatures are high enough to melt the ice but low enough to support the growth of moss, algae and other organisms.

LIFE WITHIN LIFE

The organisms and environments so far described in this chapter are largely free living. Many organisms, however, are not free living but live within, or in close association with, another organism. Such an association is called a symbiosis (which means 'living together'). The term 'symbiosis' does not imply any harm or benefit to either of the partners in the association. Where both partners gain some benefit from the association, it is called 'mutualism', and 'parasitism' where one partner lives at the expense of the other. The larger organism in a symbiotic association is called the 'host' and the smaller the 'parasite' or 'symbiont'. Being parasitised is not an unusual situation. Most animals contain parasites, usually of several types. Most groups of organisms have representatives which are parasites or symbionts. Some groups of microorganisms consist exclusively of parasites (for example, viruses), as do some phyla of animals (for example, the spiny-headed worms or Acanthocephala). There may well be more parasitic than free-living animals on Earth.

The environment of many parasitic animals consists of the inside of another animal. This means they are faced with some unusual challenges compared with a free-living species. The intestine is the most common site for parasitism, so let us look at the features of the inside of the intestine of a human as a place to live. It is dark and the intestine undergoes constant muscular movement to keep the food stirred, to aid digestion and to assist its passage through the gut. This, and the flow of food, means that a parasite is in constant danger of losing its footing and being swept out of the body. Many parasites have various types of attachment organs, consisting of a variety of suckers, clamps and hooks, which enable them to hang on to the wall of the intestine and to maintain their position. The inside of the intestine is low in oxygen and parasites, at least in some parts of the gut, have to cope with anoxic conditions. The stomach is very acidic, due to the secretion of hydrochloric acid which helps break up the food. Intestinal parasites have to survive passage through the acid conditions of the stomach and some even live there. The host is continually secreting enzymes into the intestine, to digest the food, which the parasite has to neutralise in order to prevent them from dissolving its own tissues. On the plus side, the parasite is protected from the external environment in relatively constant conditions and is provided with a ready supply of food by its host.

A mutualistic symbiosis operates to the benefit of both partners. We have already met some of these. The tube worms living around deep-sea hydrothermal vents have a mutualistic association with chemotrophic bacteria. The tube worms harbour the bacteria in their tissues and supply them with the oxygen and sulphides they need via the blood. The bacteria are able to live in a favourable position close to the vent since they are within the body of the tube worm which is anchored to the seafloor. This prevents them from being swept away. In return, the tube worms receive the products of the bacteria's metabolism as food, without which they would be unable to live in the hydrothermal vent habitat. You could say that the tube worm has acquired the ability to utilise a novel food source (the sulphides from the vents) by forming a mutualistic relationship with the bacterium. Angela Douglas of the University of York suggests that we might even regard mutualistic symbiosis as an alternative mechanism of evolution. Rather than acquiring a new ability through natural selection operating on random mutations, the tube worm has acquired the ability to utilise sulphides as a food source by forming an association with a chemotrophic bacterium. There are other examples. Many invertebrate animals contain algae. They have acquired the ability to harvest the energy in sunlight by forming a mutualistic association with a photosynthetic organism. Animals do not have the ability to digest cellulose, the material which makes up the cell wall of plants, on their own. A cow can feed on grass because there are bacteria and protozoa within some of the chambers of its stomach which produce cellulase, the enzyme necessary to digest cellulose. Many plant-eating animals have acquired the ability to digest cellulose via a mutualistic association with microorganisms. It is now widely accepted that animals and plants themselves developed via mutualistic associations between microorganisms. Mitochondria (the energy-producing structures of plant and animal cells) developed from bacterial symbionts and chloroplasts (the sites of photosynthesis in plant cells) developed from a mutualistic association with a photosynthetic microorganism.

THE FEATURES OF EXTREME ENVIRONMENTS

In this chapter, we have seen that there are a wide variety of extreme environments on Earth. They are considered to be extreme for a variety of reasons, including high or low temperatures, lack of water, high or low pH, high pressures, high exposure to radiation (particularly ultraviolet), high salt concentrations, exposure to toxins and low nutrient availability. These stresses rarely act on their own and organisms are exposed to a combination of stresses (Table 2.1). We can divide them into two broad groups: terrestrial and aquatic (including salt lakes, soda lakes and hot springs). Water availability is a major factor in terrestrial habitats, but not, of course, in aquatic habitats. The great bulk of water in most aquatic habitats also has a buffering effect and so changes in

Table 2.1 *Features of extreme environments*

Habitat	Temperature	Water	Pressure	Oxygen	pH	Toxins	Nutrients	Salts	Ultraviolet radiation
Hot deserts	←	→	—	—	—	—	—	—	←
Cold deserts	↑↓	→	—	—	—	—	—	—	←
Temporary deserts	↑↓	→	—	—	—	—	—	—	—
Salt lakes	—	—	—	—	←	←	←	←	→
Soda lakes	—	—	—	—	←	←	←	←	→
Polar regions	→	→	—	→	—	—	—	↓↑	←
Mountains	↑↓	→	—	—	—	—	—	—	←
Temperate winter	→	→	—	—	—	—	—	—	—
Deep sea	—	—	←	—	—	—	→	—	→
Hydrothermal vents	←	—	←	→	→	←	←	←	→
Cold seeps	—	—	←	—	—	←	←	—	→
Deep subsurface	←	—	←	—	⇌	←	→	←	→
Hot springs	←	—	—	—	—	←	←	←	—
Parasitic	—	—	—	→	→	—	←	—	→

Notes:
↑ is high, ↓ is low and — is normal

temperature, and other conditions, are likely to be slow. We might say that, overall, extreme terrestrial habitats are more extreme than extreme aquatic habitats.

Some extreme environments are constantly extreme. This favours organisms with capacity adaptations which can grow and reproduce under the extreme conditions. Examples include: the deep sea (high pressure), polar oceans (freezing temperatures), and hydrothermal vents and hot springs (high temperatures). Environments where extreme conditions are temporary and there are periods of less extreme conditions when growth and reproduction can occur may tend to favour resistance adaptation, at least in some organisms. Examples include: deserts, polar regions and mountains. Again, this reveals a difference between aquatic and terrestrial environments.

The following chapters will look at how some of the organisms which live in extreme environments survive the stresses to which they are exposed.

3 Life without water

As a child, I lived in the Midlands in the centre of England. Family holidays meant a long car journey and I can still remember the excitement of that first rare glimpse of the sea. We feel a similar sense of elation on discovering a waterfall or a lake during a walk in the country or on hearing the patter of rain after a long dry period. Perhaps the emotional response we feel to water is a recognition of how important it is to us and to life in general. For we all know that water is essential for life. However, although it is true that life cannot exist without at least periodic access to water, organisms can live in some very dry places and some can survive the almost total loss of water from their bodies.

Why is water so important? Water comprises by far the largest proportion of the chemical make-up of the bodies of most organisms, usually from 60 to 90 per cent. Indeed, as JBS Haldane once remarked, 'Even the Pope is 70 per cent water'. Living organisms need metabolic processes which maintain their structure and enable them to move, grow and reproduce. Metabolism involves chemical reactions and these reactions need a medium in which they can take place in a controlled manner. Consider, for example, how organisms get their energy. Energy comes mainly from the oxidation or burning of sugars. It requires a sugar (such as glucose or sucrose) and oxygen. However, if you leave a bowl of sugar in contact with air, it does not spontaneously oxidise (or only does so very slowly) or burst into flames. If you throw sugar onto a fire, it burns violently, undergoing an oxidation which releases its chemical energy in the form of heat and light. When dissolved in water within the cells of an organism, however, the sugar can be oxidised in a controlled manner by a series of reactions which harvest the energy within its chemical structure. The energy can then be stored and utilised by the cell. Without water, the chemical reactions involved cannot occur and there will be no metabolism and no manifestation of life. As we will see later, water is also an important part of the structure of membranes and of proteins, nucleic acids and other biological molecules. It thus contributes to the structural order of cells.

LIFE RUNNING SHORT OF WATER

In Chapter 2, we saw that there were a number of terrestrial environments in which the availability of water was likely to be a problem. Organisms have two main responses to situations where water is in short supply and is only sporadically available. The first is a suite of capacity adaptations. The organism taps the sources of water that are available, such as the tenebrionid beetles which collect moisture from the fog that forms in the Namib Desert or plants which send out long roots to seek water. Some plants, such as cacti, store water to use during dry periods. Plants and animals have adaptations which conserve the water within their tissues. They have a waxy cuticle or skin and a covering of hairs or spines to restrict water loss. Animals may produce a concentrated urine and dry faeces and recover water from their breath. The second response to a lack of water is to lie dormant until water becomes available again.

LIFE WITHOUT WATER

Most animals and plants have only a limited ability to survive water loss. Humans may die if they lose 14 per cent of the water from their bodies. Some frogs can lose 50 per cent and some earthworms 83 per cent of their water and still recover. Some organisms, however, can lose more than 95 per cent, or even more than 99 per cent, of their water and enter into a state of anhydrobiosis (life without water) in which their metabolism comes, reversibly, to a standstill. There is a problem, however, in defining which organisms are capable of anhydrobiosis. Organisms show a whole range of abilities to survive water loss, ranging from losing just a little bit (as can humans) to losing almost all of it. At what point do we consider an organism to be anhydrobiotic?

Anhydrobiosis is best defined in terms of its effect on metabolism. Losing a little bit of water may have no effect on metabolism. Losing more may depress metabolism, but further water loss will cause it to cease altogether. Consider what happens to metabolism in mammals which survive the winter in a state of deep hibernation. Hibernating mammals have a reduced rate of metabolism, but they do not cease metabolising altogether. We have one term for animals which have ceased metabolising – we refer to them as being 'dead'. The difference between anhydrobiosis and death is that an organism in a state of anhydrobiosis will recover, grow and reproduce when normal conditions return (when immersed in water), whereas a dead organism will not. We could thus define anhydrobiosis as 'the ability to survive the cessation of metabolism due to water loss'.

Anhydrobiosis represents an extreme example of a resistance adaptation, where the organism shuts down operations completely as a result of water loss and survives in an ametabolic state until water returns. Are there any organisms which can continue to function in the absence of water? The answer is almost certainly no, given the vital role of water in metabolism. There have been, however, some interesting reports concerning nematodes from the Dry Valleys of Antarctica. Diana Wall, from Colorado State University, and her coworkers, have reported the presence of nematodes in the very dry soils of the Dry Valleys. These must be relying on very occasional inputs of water from melting snow or from glaciers, or are blown there from more productive sites. They presumably spend most of their time in anhydrobiosis.

ANHYDROBIOTIC ORGANISMS

Among animals, there appears to be an upper limit on the size and complexity of those which are capable of anhydrobiosis. The largest such animal is the midge larva *Polypedilium vanderplanki* from rain-filled rock pools in Africa, which is about 5 millimetres long (Figure 3.1). *P. vanderplanki* and the larvae of a few other species of chironomids or midges are the only higher insects known to survive anhydrobiotically. Springtails (collembolans) are a group of primitive wingless insects, a

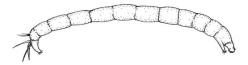

FIGURE 3.1 The larva of the midge *Polypedilium vanderplanki,* found in rain-filled ponds in Africa, is the largest known anhydrobiotic animal (about half a centimetre long). Drawing by Jo Ogier (redrawn from Hinton, 1960).

few species of which are capable of anhydrobiosis. A number of crustaceans which inhabit temporary and saline ponds can survive anhydrobiotically; the best known of these is the brine shrimp *Artemia,* which survives as an encysted embryo. Nematodes, rotifers and tardigrades are groups of invertebrates, consisting mainly of species which are microscopic in size. They need at least a film of water for activity and growth, but, in some environments, they are exposed to desiccation for varying periods of time. Such environments include soil, moss, deserts, temporary ponds, terrestrial polar habitats, the aerial parts of plants and, for plant-parasitic nematodes, plant tissue which dries out when the plant dies or sets seed. Not all species of these invertebrates are capable of anhydrobiosis; their desiccation survival abilities match the stresses they face in their environment.

Most plants have stages in their life cycles which are very dry. Dehydration is part of the maturation process of plant seeds and most have a water content of 5–20 per cent. Some have even lower water contents; for example, the seeds of birch trees have a water content of only 0.01–0.4 per cent. At such low water contents, these seeds are likely to be anhydrobiotic. The pollen (male reproductive agent) of many plants is also very dry and can tolerate the desiccation experienced during its dispersal by the wind or on the bodies of insects. Most plants lose the ability to survive desiccation once their seed germinates and starts to grow. There are a few plants, however, that can survive anhydrobiosis in their mature growth forms. Some plants from southern Africa can survive after losing so much water that their leaves crumble into dust if rubbed between the fingers. These 'resurrection plants' will survive exposure to 0 per cent relative humidity and a water content of less than 5 per cent and yet recover and grow when it rains. Over 100 species of resurrection plants have been described, mostly from the hot dry regions of Southern Africa and Australia, including species from a variety of groups of flowering plants and from ferns and their allies. These plants are the first to colonise rock surfaces and shallow soils, where they are likely to be exposed to extreme desiccation and conditions which are too harsh for other plants to survive.

Mosses have no roots and are dependent on absorbing water across their surface, which confines their growth to habitats which are, at least periodically, wet. Nevertheless, many species can survive desiccation. This is particularly important for mosses that live in dry regions such as deserts, but also for those that live on the surface of rocks or of plants (such as the bark of trees), where they may occasionally dry out. The spores of mosses, and of other groups of primitive plants, can also tolerate desiccation. A wide variety of algae are associated with sites where they are exposed to desiccation, such as the surface and interior of rocks and in desert and other arid soils. These must also have life cycle stages which can survive anhydrobiotically. Algae (and cyanobacteria) also associate with fungi to form lichens which colonise desiccation-prone sites, including bare soil, the surface of rocks and tree trunks. Many fungi can tolerate desiccation and their spores are particularly resistant. Yeasts, which are single-celled fungi, will survive dehydration if desiccated and rehydrated under the right conditions. Yeasts, which are used for making bread, wine and beer, are often supplied commercially as a dry powder. Protozoa can also survive anhydrobiosis, particularly as cysts.

Many bacteria have some ability to survive anhydrobiosis, although their ability to do so depends on the rate of desiccation. Some will survive high rates of water loss, while others require a slow rate of water loss in order to survive. In general, however, bacteria can survive better after slow rather than fast drying. Spores are particularly resistant. Cyanobacteria are prominent members of the bacterial communities of a variety of extreme environments. They survive anhydrobiotically in deserts, polar regions and in a variety of other terrestrial sites where they are exposed to desiccation, such as depressions in rocks and the surface of roofs.

Although this may seem already to be a fairly long list of organisms, anhydrobiosis is probably much more widespread than we presently realise. As we will see later, many anhydrobiotic organisms need a slow rate of water loss in order to survive. Many anhydrobiotic animals and plants were discovered from material collected dry in the field, where they naturally experience slow water loss. Drying an organism on a glass slide on the laboratory bench or over a desiccant is not a fair test of its ability to survive anhydrobiosis. It needs to be dried at a rate which mimics the rates of water loss it is likely to experience in its natural environment. For many, this means drying them very slowly indeed. Using environmentally relevant rates of desiccation will certainly allow us to discover many more anhydrobiotic organisms.

A LITTLE DRY HISTORY

The history of the study of anhydrobiosis is not a dry topic at all, but an intriguing story involving professional jealousies and matters of life and death. The early history has been told in some detail by David Keilin in his Leeuwenhoek lecture, given to the Royal Society of London in 1958, and much of the following draws on his account. The Leeuwenhoek lecture honours the Dutch scientist Antoni Van Leeuwenhoek, one of the first microscopists. Using his microscopes, Leeuwenhoek made many discoveries, including the first observation of bacteria, and he is regarded as the father of microbiology. His observations on anhydrobiosis were made rather late in his life (in 1702, aged 70) and were recorded in his letter to the Royal Society 'On certain animalcules found in the sediments in gutters of the roofs of houses'. Here is his description of what he found:

> I have often placed the Animalcules I have before described out of the water, not leaving the quantity of a grain of sand adjoining to them, in order to see whether, when all the water about them was evaporated and they were exposed to air, their bodies would burst, as I had often

seen in other Animalcules. But now I found that when almost all the water was evaporated, so that the creature could no longer be covered with water, nor move itself as usual, it then contracted itself into an oval figure, and in that state remained, nor could I perceive that the moisture evaporated from its body, for it preserved its oval and round shape unhurt.

In order more fully to satisfy myself in this respect, on the third of September, about seven in the morning, I took some of this dry sediment, which I had taken out of the leaden gutter and had stood almost two days in my study, and put a little of it into two separate glass tubes, wherein I poured some rain water which had been boiled and afterwards cooled...

As soon as I had poured on the water, I stirred the whole about, that the sediment which, by means of the hairs in it, seemed to adhere like a solid body, might the sooner be mixed with the water: and when it had settled to the bottom of the glass, I examined it, and perceived some of the Animalcules lying closely heaped together. In a short time afterwards they began to extend their bodies, and in half an hour at least a hundred of them were swimming about the glass, though the whole of the sediment which I had put into it did not, in my judgement, exceed the weight of two grains...

The preceding kinds of experiment I have many times repeated with the same success, and in particular with some of this sediment, which had been kept in my study above five months, and upon pouring on it rain water, which had been boiled, and afterwards cooled, I saw in a few hours' time many of the Animalcules before described. And if, after being so long in a dry state, these Animalcules, upon water being given to them can unfold their bodies and move about in the usual manner, we may conclude, that in many places, where in summer time the waters stagnate, and at length dry up, there may be many kinds of Animalcules, which, though not originally in those waters, may be carried thither by water fowl, in the water or mud adhering to their feet or feathers.

There are a number of points we might note from this account. The little animals ('Animalcules') were recovered from dry sediment that had been stored for five months, indicating that they could survive desiccation for at least this long. There was a change in the shape of the animals, both during desiccation and during rehydration, and there was a delay in the recovery of activity after rehydration. These are features which have been observed in more recent studies. The 'Animalcules' are believed to have been rotifers, although such sediments are likely to contain other microscopic animals such as tardigrades and nematodes, as well as protozoa and other microbes.

As Keilin notes, these observations by Leeuwenhoek apparently did not excite the interest of his contemporaries, who may have found the microscopic world which he described a little too outlandish. It was not until 40 years later that the next observation of anhydrobiosis was made, by John Needham, in 1743. Needham was interested in a disease of wheat which results in the formation of galls, known as 'cockles' or 'peppercorns'. Needham describes how, when he opened an infected wheat grain, he found a soft white substance which consisted of fibres. When he added water, the previously dry and lifeless fibres separated and started moving. The fibres were in fact nematodes. Needham's account was not only the first description of anhydrobiosis in nematodes but also the first account of a plant-parasitic nematode. The nematode was probably *Anguina tritici* which accumulates in infected wheat as an infective larva in the grains, transforming them into galls. *Anguina* is one of the most desiccation tolerant of plant-parasitic nematodes.

Soon after (in 1753), Needham's observations were confirmed by Henry Baker, the author of several books on microscopy. Baker not only confirmed Needham's and Leeuwenhoek's observations but made a rather extraordinary claim:

> We find an Instance here, that *Life* may be suspended and seemingly destroyed; that by an Exhalation of the Fluids necessary to a living Animal, the Circulations may cease, all the Organs and Vessels of the Body may be shrunk up, dried and hardened; and yet, after a long while, Life may begin anew to actuate the same Body; and all the animal Motions and Faculties may be restored, merely by replenishing the Organs and Vessels with a fresh supply of Fluid.

This was the first claim that life could be a discontinuous process and that an animal could apparently 'die' after desiccation, showing no signs of life, and yet 'come back to life again' when water was added. This claim of death and resurrection had obvious religious implications, with several religions, not only Christianity, including such a process as part of their tradition. The idea that death and resurrection occurs in nature, even among lowly organisms, is thus a controversial one.

Lazzaro Spallanzani, Professor of Natural History at the University of Paris and one of the leading scientists of his time, initially denied the animal nature of the dried fibres observed by Needham in wheat grains (in his book published in 1767). Spallanzani considered them to be dried vegetable fibres, with the movement resulting from water penetrating into them. The reputation of Spallanzani was such that it led Needham to recant his view. Shortly after this, however, two scientists, l'Abbé Roffredi and Felice Fontana, working independently, described the life cycle of the nematode responsible for the formation of galls in wheat and confirmed the animal nature of the fibres. Roffredi and Fontana in their publications attacked Needham for withdrawing his earlier correct view. Needham accepted their findings, but was rather hurt by their criticism. He complained to the editor of the journal which published their papers about instances 'where the critic aims only to wound his adversary'.

Following the work of Roffredi and Fontana, Spallanzani re-examined the phenomenon and was able to confirm the work of Leeuwenhoek, Needham and Baker for himself and to extend their studies greatly. He showed that dried rotifers were more resistant to high temperatures ($73\,°C$) than were hydrated rotifers (which died at $45\,°C$), but that both dried and hydrated rotifers could survive freezing to $-24\,°C$. Dried rotifers could also survive exposure to a vacuum. His experiments led him to accept the idea of death and resurrection in these animals, stating in a paper entitled, 'Observations and experiments upon some singular animals which may be killed and revived':

> An animal, which revives after death, and which within certain limits, revives as often as we please, is a phenomenon, as incredible as it seems improbable and paradoxical. It confounds the most accepted ideas of animality; it creates new ideas, and becomes an object no less interesting to the researches of the naturalist than to the speculation of the profound metaphysician.

Not everyone accepted Spallanzani's view. One of the strongest opponents of the idea that animals underwent death and resurrection during desiccation and rehydration was Christian Ehrenberg who felt that life processes were greatly slowed down and not stopped altogether during anhydrobiosis and also that not all water was lost. The controversy erupted in 1858 in a debate between two French scientists, PLN Doyère and Félix-Archimède Pouchet, both of whom conducted experiments on tardigrades and rotifers but reached very different conclusions from their work. Doyère considered that these animals could be revived after complete desiccation and the cessation of their life processes, while Pouchet thought that no organism could survive complete desiccation or return to life after all life processes had stopped. The debate became so heated that members of the learned societies and even the newspapers of Paris became divided into two fiercely opposed groups: the resurrectionists and the anti-resurrectionists. In 1859, Doyère and Pouchet approached the Biological Society of France asking them to give an impartial ruling. The Society established a special commission which examined the work of the two scientists and conducted experiments of their own. An extensive report was written by Paul Broca, a distinguished French anatomist, which generally supported the views of Doyère.

THE ANHYDROBIOTIC STATE
Further defining anhydrobiosis
Broca's report quieted the controversy for a while but it has resurfaced periodically ever since and continues today. The main two points at issue are: do anhydrobiotic organisms lose all their water during desiccation and does metabolism (which we nowadays think of as the

primary manifestation of 'life processes') cease? It should be noted that both these questions are essentially unanswerable. If no water is detected in a desiccated anhydrobiotic organism using a particular technique, there might still be some water present which the technique cannot detect and, if no metabolism can be detected, there may be a level of metabolism occurring which the technique used is not sensitive enough to reveal. This latter point was addressed by John Barrett at the University of Wales. He suggested in 1982 that, while we could not prove that metabolism had ceased, we could say what level of metabolism a particular technique should be able to detect. He used three different techniques to look for metabolism in the anhydrobiotic nematode *Ditylenchus dipsaci* (a parasite of plants): oxygen uptake, heat output and the production of carbon dioxide. No metabolism was detected in the dry nematodes using any of these techniques. Oxygen uptake should have detected a metabolic rate of 0.3 per cent of normal levels, heat output a rate of 0.06 per cent and carbon dioxide a rate of 0.01 per cent. So, if there is metabolism occurring in the desiccated nematodes, it is at a rate which is less than 1/10000 of that of hydrated nematodes. It is perhaps not unreasonable to accept that metabolism had ceased altogether.

It may not be necessary for an organism to lose all, or nearly all, its water for metabolism to cease and for it to become anhydrobiotic. James Clegg, now at the University of California's Bodega Marine Laboratory, from his studies on the cysts of the brine shrimp *Artemia*, proposed in the 1970s that, as an organism loses water, it passes through a number of metabolic states. When it has 40–100 per cent of its normal water content, this is sufficient to form a continuous bulk of water within the cells of the organism. This provides a medium for the series of chemical reactions involved in the metabolism of the organism, allowing it to proceed as normal. At 20–40 per cent water, there is insufficient for it to form a continuous medium, only isolated reactions in the remaining pockets of water are possible, most sequences of reactions become impossible and metabolism is depressed. Below 20 per cent water, there is no free water within cells and metabolism may cease (and, if the organism survives, it is anhydrobiotic). The water which remains is thought to be tightly bound to, or in close association with, the proteins and other substances which make up the structure of the cell. This is referred to as 'bound water', 'unfreezable water' (since about 20 per cent of cell water cannot be easily frozen) or 'osmotically inactive water' (since it is not free to move under an osmotic stress, although it will move along gradients of water potential such as evaporation under desiccating conditions). A somewhat smaller quantity of water is actually involved in the structure of biological molecules (0.15–0.4 per cent). The water content of dried cells in anhydrobiosis is too low to provide even a single layer of water molecules to coat their internal surfaces. Under such circumstances, the usual sequence of metabolic reactions is clearly impossible. As well as reducing the availability of water, desiccation will also increase the concentrations of dissolved organic and inorganic substances within the cells. Metabolic enzymes can only tolerate a limited range of concentrations of some of these substances and it is likely that an increase in their levels during desiccation will result in metabolism being inhibited.

Whatever the mechanism, it is clear that normal metabolism ceases when a substantial proportion of water is lost, but long before it is lost altogether. If we define anhydrobiosis as 'the ability to survive the cessation of metabolism due to water loss', it is clear that the organism does not need to lose all its water for this to occur. Since it may be difficult to demonstrate that metabolism has ceased, we can consider an organism to be in a state of anhydrobiosis if it survives the loss of more than 80 per cent of its water (since there is no free water present in cells and normal metabolism ceases at these water contents). The ability to survive the loss of some osmotically inactive water appears to be critical for anhydrobiosis. Water will be lost from an organism whenever the water activity outside its body is lower than that inside its body – that is, when it is drier outside than it is inside. The water will continue to be lost until the water activities inside and outside the body become equal. This equalisation of the water activities may occur by water being lost from the organism or by it producing substances (osmotically active solutes) that may replace some of the water bonded to proteins, and other large molecules, within its cells. Desiccation-sensitive bacterial cells die if their water content is reduced to about 6 per cent and this occurs if they are dried at a relative humidity of 80 per cent or below. At such low water contents, metabolism ceases. Organisms will need the mechanisms which enable them to survive anhydrobiotically (if they can) when they are faced with these sorts of relative humidities, unless they can prevent water loss from occurring. The rate of water loss, and hence the time available to the organism to make any necessary adjustments, will depend on how dry the air surrounding it is.

Anhydrobiosis and the nature of life

When metabolism ceases during anhydrobiosis, so do all the processes that we usually consider to be a part of life. I remember that at school I was taught a list of the characteristics of living things which distinguished them from non-living things. The list went something like this: living things move, respire, feed, excrete, grow, reproduce and respond to stimuli. To be considered alive, an object had to exhibit all these characteristics. A car moves and might be considered to feed (on petrol) and excrete (exhaust fumes), but we never see a baby car emerging from its exhaust pipe. What then are we to make of anhydrobiotes which, in the dry state, exhibit none of these characteristics. Should they be considered dead or non-living when in this state? Clearly, they have the capacity for life, since when immersed in water they recover and display the characteristics we associate with life. In anhydrobiosis, life exists in a purely structural state. We can thus recognise three states for an organism: alive, dead and anhydrobiotic (or cryptobiotic, which includes the cessation of metabolism in response to stresses other than desiccation). Perhaps the key feature of living organisms is the potential for reproduction – the presence of a self-replicating molecule, DNA or RNA, which carries the blueprint (genetic code) for their construction and which is the product of a process of evolution.

The ageing process is suspended while an organism is dry. Many nematodes complete their life cycle and die within a matter of weeks of hatching from the egg. They will, however, survive in a state of anhydrobiosis for many years. We thus need to distinguish between the chronological and physiological age of an organism. There are many records of plant parasitic nematodes surviving anhydrobiotically for decades. In tardigrades, the record is 120 years, in animals recovered from a dried plant stored in an herbarium. The proportion of organisms which recovers when stored dry, however, slowly declines. What causes this slow death rate among organisms in an anhydrobiotic state? Many survive better if stored in an atmosphere of nitrogen gas, rather than air. Some biological molecules react with the oxygen in the air (oxidise) and this is often destructive. Storing the material in the dark and at low temperature also improves survival.

A living organism continually repairs any damage which occurs to its cells. During anhydrobiosis, normal metabolism ceases and this repair cannot occur. Any damage to the cells of the organism thus accumulates and may reach a point where death occurs when it is reimmersed in water. Just as life processes are suspended during anhydrobiosis, so are some of the processes which lead to death. Howard Hinton, formerly Professor of Entomology at the University of Bristol, working on anhydrobiotic larvae of the midge *P. vanderplanki*, reported that a larva which had suffered damage to its body wall while dry had a portion of its gut forced through the wound when it was immersed in water; it recovered activity only to die some four hours later.

Anhydrobiosis and extreme survival
In the dry state, anhydrobiotes are resistant to physical insults which would be fatal to them if they were hydrated. Since there is no or little water present to freeze, they are resistant to low temperatures. Survival to temperatures within a fraction of a degree of absolute zero ($-273\,°C$) has been reported. At the other extreme, anhydrobiotic nematodes have survived two minutes at $105\,°C$, tardigrades and rotifers

several minutes at 151 °C, *Artemia* cysts an hour and a half at 103.5 °C and *P. vanderplanki* larvae a minute at 102–104 °C. Anhydrobiotic animals have been reported to survive high doses of X-ray radiation and immersion in alcohol, and anhydrobiotic nematodes can tolerate the nematocides designed to kill them. Anhydrobiotic nematodes have survived exposure to vacuums and tardigrades have survived high pressures of 6000 atmospheres. Some of these reports must be treated with caution, however, because what may have been observed was delayed death rather than survival. Tardigrades have revived after exposure to the vacuum and high doses of radiation inside the column of a scanning electron microscope, only to die after a few minutes of movement.

The remarkable survival abilities of anhydrobiotic animals have led some to suggest that they may be able to survive conditions in space. Reinhardt Kristensen of the Zoological Museum in Copenhagen is quoted as saying with regards to tardigrades: 'They can tolerate outer space, no doubt about it'. The survival abilities of *Ditylenchus dipsaci*, a plant-parasitic nematode, led John Barrett of the University of Wales and myself to wonder whether they might survive exposure to space. In what we called (among ourselves) the 'Worms in Space Project' in the mid-1980s, we booked them on the European Retrievable Carrier (EURECA) due to be launched by the European Space Agency on the Space Shuttle. However, although *Ditylenchus* can survive exposure to a vacuum, it failed to survive in the ultra-high vacuum facility operated by the German Aerospace Centre in Cologne which simulates the vacuum of space. Perhaps they can not survive losing that last drop of water or some other critical material is lost in extreme vacuums. I am afraid that our worms lost their seat on the Space Shuttle and our experiments have remained earth bound.

Although the survival abilities of anhydrobiotic animals in space have yet to be tested, some microorganisms have made the journey into the void. In 1984, the Long Duration Exposure Facility (LDEF) was launched by the National Aeronautics and Space Administration (NASA) on board the Space Shuttle Challenger and placed in low Earth orbit at an altitude of 476 kilometres above the Earth. The original intention was that the LDEF be retrieved after a year, but the 1986 Challenger disaster, in which the launch rocket exploded during take-off killing all seven crew members, and other problems with the Space Shuttle programme meant that it was not recovered until 1990, after 69 months in orbit. Of the biological samples on board the LDEF, only bacterial spores were exposed to the environment of space. Other material included *Artemia* (brine shrimp) cysts and plant seeds, but these were shielded from the vacuum of space within sealed containers. Bacterial spores (*Bacillus subtilis*), however, were exposed to space for nearly six years. Even some unprotected spores survived after their return to Earth (about 2 per cent), but survival was greater (up to 70 per cent) if they were protected by buffer salts, glucose or by an aluminium cover which shielded them from ultraviolet (UV) radiation. Since then, bacterial spores have flown on other missions which have exposed them, along with other microbes, to the space environment. Fungal spores, viruses and salt-tolerant bacteria and cyanobacteria, as well as bacterial spores, have all survived exposure to space.

Damage to the organisms' DNA, caused by exposure to the vacuum and radiation of space, appears to be important in inducing mutations and in determining their survival. UV radiation from the sun causes the most damage and is directly absorbed by the DNA molecule. It reacts with the DNA to produce products which are highly lethal and mutagenic. Microorganisms can, however, be protected from UV radiation by dust particles or other substances which shield them from it. Even a multiple layer of bacterial cells will protect those at the bottom of the layer from the harmful effects of UV radiation. Organisms have so far been exposed to space only in low Earth orbit. Temperatures during these exposures varied from $-30\,°C$ to $+45\,°C$ due to direct solar heating or to heat reflected from the Earth. Temperatures in deep space are much lower $(-269\,°C)$, within a few degrees of absolute zero. At such low temperatures, the harmful effects of UV radiation are greatly reduced.

Among the cosmic radiation, heavy high-energy (HZE) particles cause the most damage to biological materials. These particles are of galactic rather than solar origin and are atomic nuclei which are stripped of their electrons and accelerated to high speeds. These particles are fairly rare (about 1 per cent of the particulate radiation in space), but a single hit by a HZE particle can kill a spore and the chances of such a hit will set the ultimate time limit on their survival in space.

THE MECHANISMS OF ANHYDROBIOSIS IN ANIMALS

How do anhydrobiotic animals survive in their remarkable way? Most of the information we have is from studies on nematodes and tardigrades, so I will focus on these groups but add results of studies on other organisms where appropriate.

Controlling water loss

A slow rate of water loss appears to be important to prevent damage to the structure of the animal and to allow it to make biochemical changes in preparation for the desiccated state. Anhydrobiotic nematodes are divided into two broad groups: slow-dehydration strategists and fast-dehydration strategists. Slow-dehydration strategists rely on the characteristics of their environment to produce the slow rate of water loss needed. For a nematode living in soil or moss, the material in which it lives will lose water slowly when it is exposed to desiccation. This will produce the slow rate of water loss necessary for the nematode's survival. Some nematodes, however, live in more exposed sites such as the aerial parts of plants and drying plant tissue which will lose water quickly when exposed to desiccation (Figure 3.2). In these habitats, the nematode itself needs to control the rate at which water is lost from its body.

Nematodes are covered by a cuticle which, in many cases, has a very low permeability to water, slowing the rate at which they lose water. Some appear to decrease the permeability of their cuticles when they are exposed to desiccation. This also happens in tardigrades, where the secretion of lipids through pores in the cuticle and changes in the properties of the cuticle itself reduce water loss. A similar job is performed

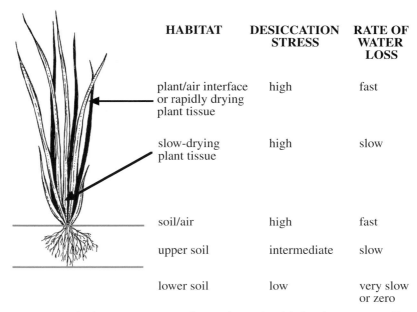

	HABITAT	DESICCATION STRESS	RATE OF WATER LOSS
	plant/air interface or rapidly drying plant tissue	high	fast
	slow-drying plant tissue	high	slow
	soil/air	high	fast
	upper soil	intermediate	slow
	lower soil	low	very slow or zero

FIGURE 3.2 The desiccation stress and rates of water loss likely to be experienced by plant-parasitic and free-living nematodes in various plant and soil habitats. Based on a figure in Womersley (1987). Drawing of plant by Jo Ogier.

by the cuticles of other animals and of plants, by eggshells, the cases of seeds and the walls of cysts and spores. Some algae, and other microorganisms, secrete a gelatinous material which slows the rate of water loss. Changes in behaviour may also act to reduce the rate of water loss. Many nematodes coil up when desiccated (Figure 3.3). This reduces the surface area exposed to the air and consequently reduces the rate of water loss. Tardigrades withdraw their legs into their bodies forming a tun (so-called because it is barrel shaped). This reduces the area of exposed surface but may also slow water loss by removing the more permeable areas of the tardigrade's cuticle from exposure to the air (Figure 3.4). Some plant-parasitic nematodes form dense aggregations known as 'eelworm wool'. The formation of these aggregations may

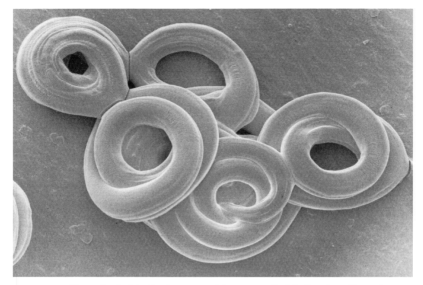

FIGURE 3.3 The anhydrobiotic plant-parasitic nematode *Ditylenchus dipsaci*, coiled in response to desiccation. Each worm is about 1 millimetre long.

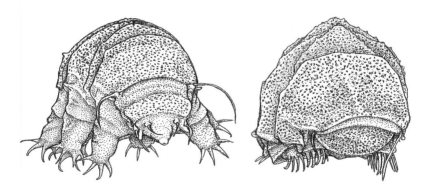

FIGURE 3.4 An active tardigrade (left) and one which has withdrawn its appendages into its body to form a tun during desiccation and entry into anhydrobiosis (right). The animal is about 1 millimetre long. Drawing by Jo Ogier, based on photographs in Crowe & Cooper (1971).

aid desiccation survival since the nematodes on the outside will dry first and slow the rate of water loss of those in the centre of the aggregation (the so-called 'eggshell effect').

In the dry state

A slow rate of water loss allows the animal to shrink and to pack its internal structures together in an orderly fashion so that disruption during desiccation is prevented or at least minimised. The structure of anhydrobiotic organisms in the dry state can be revealed using techniques for preparing specimens for electron microscopy which do not involve exposure to water. Studies on a wide range of anhydrobiotic organisms from cyanobacteria to nematodes reveal a similar picture. Most anhydrobiotes keep their internal structures intact in the dry state. The main change during desiccation is that the cytoplasm of the cell is reduced in volume and becomes condensed around the mitochondria and other cell structures. Organisms in a state of anhydrobiosis have much the same appearance as they do in the active hydrated state. The main difference is that the cells shrink and the cytoplasm and organelles pack closely together. Some more complex changes have been observed in anhydrobiotic plant tissues. Multilayered structures and dense spheres, which are absent in wet tissue, have been seen in some plant pollen. The origin and function of these structures is unknown.

You might think that desiccation would harm the enzymes, DNA and other macromolecules which make up the bodies of organisms. However, enzymes show no loss of activity in anhydrobiotic nematodes and there is also no evidence of the break up of DNA molecules.

The role of trehalose

Trehalose is a sugar which is closely related to the sugar (sucrose) which you put in your tea. A number of anhydrobiotes, including nematodes, tardigrades, *Artemia* and yeast, produce trehalose as they dry out. Anhydrobiosis is thus the sweet life as well as the dry life. John Crowe of the University of California at Davis has been the main proponent of the theory that trehalose plays a crucial role in anhydrobiosis. He found that *Aphelenchus avenae*, a fungus-eating nematode which could be cultured in large quantities for biochemical analysis, converted its carbohydrate (sugar) stores from glycogen (the main store of carbohydrates in animals) into trehalose as they dried out. James Clegg found a similar phenomenon in *Artemia* cysts. This increase in trehalose levels led these researchers to suggest that trehalose plays an important role in anhydrobiosis. They also found that anhydrobiotic organisms converted lipid into glycerol (a polyol or sugar alcohol) and suggested that the glycerol replaced the bound water which makes up 20 per cent of the water in the body. It is now thought that the glycerol was produced in response to conditions within their samples becoming low in oxygen. Glycerol is produced in response to a variety of stresses such as low oxygen concentrations, osmotic stress and low temperatures. It thus appears to be a general response to stress and is now not thought to be specifically involved in anhydrobiosis. There is, however, strong evidence that trehalose plays an important role.

A biological membrane consists of two layers of phospholipids with proteins immersed within the bilayer and carbohydrates associated with its surface. The structure of the membrane is maintained by its interaction with water. The phospholipid molecules have two ends, one end of which is attracted to water (the hydrophilic heads) and the other is repelled by water (the hydrophobic tails). Water molecules attach to the hydrophilic heads. If you remove the water, the structure of the membrane is changed. The normal condition of a membrane is a liquid crystalline state. In this state, the membrane is fluid, with the same mobility as salad oil, and its molecules can move around. This fluidity is important for the biological functions of membranes. The condition of the membrane will, however, change as conditions within the organism change. If the temperature falls, for example, the membrane becomes less fluid and eventually solidifies, like melted butter when it cools. Desiccation also affects the condition of the membrane. Removal of water allows the phospholipid molecules to pack more closely together and the membrane undergoes a change in state from a

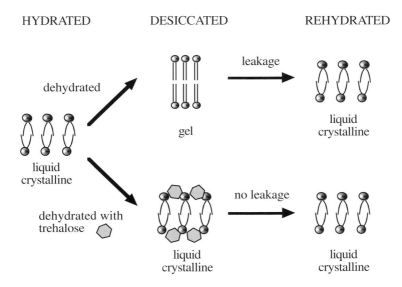

HYDRATED   DESICCATED   REHYDRATED

FIGURE 3.5 The water replacement hypothesis for the protective action of trehalose on membranes during desiccation and rehydration. Dehydration causes membranes to change from a liquid crystalline state to a gel state. Reversal of this change upon rehydration causes membranes to become transiently leaky, resulting in the fatal loss of cell contents. Trehalose replaces the water molecules associated with the membrane and prevents these harmful changes. Redrawn from a figure in Crowe *et al.* (1992).

liquid crystal to a gel. When water is added, the molecules move apart again and revert back to the liquid crystalline state. This change causes the membrane to become leaky, for a while, which results in the loss of substances from within the cell. This leakage of materials from the cell could be fatal for an anhydrobiotic organism. Trehalose attaches to the membrane and prevents it from changing from the liquid crystalline to the gel state when it loses water. When water returns, there is thus no change of state and the leakage of materials from the cell is prevented (Figure 3.5). This proposed mechanism is called 'the water replacement hypothesis' since trehalose replaces the water molecules in the membrane and prevents it from undergoing changes in state which would prove fatal during desiccation and rehydration.

Trehalose appears to play a number of other important roles in anhydrobiosis, in addition to preventing transitions in membranes which could result in the leakage of cell contents. Most sugars are susceptible to oxidation when in contact with air and form reaction products with proteins (the Maillard or browning reaction), but this is not the case with trehalose which is a non-reducing sugar. As water is lost from cells, the contents pack together and membranes which were previously separated could come into contact and fuse together. Such membrane fusion would violate the integrity of cellular compartments which would fall apart when rehydrated. Trehalose is thought to prevent membrane fusion from occurring. When you gently heat a sugar, it melts and then, on cooling, resolidifies into a glassy state. This is how boiled sweets, lollies and mints are made. Sugar glasses are also formed as a result of desiccation. As well as preventing membrane fusion, the formation of a sugar glass may trap the tissues of an anhydrobiotic organism in a sticky, immobile and stable medium which would prevent any deterioration from occurring.

Some proteins are susceptible to damage (denaturation) during desiccation; trehalose stabilises and protects these proteins. How it does so is not entirely clear. The properties of a protein depend on it maintaining its correct shape or conformation. The conformation is at least partly dependent on the coating of water molecules which bond to the surface of the protein. Between one- and three-quarters of a gram of water bind to each gram of protein in solution. This makes up most the bound water in the cell, although perhaps the term 'bound water' is misleading since it can exchange rapidly with the free water in the cell. Loss of this water, however, could affect the conformation of a protein and result in the loss of its vital biological properties (denaturation). Trehalose prevents protein denaturation during desiccation, perhaps by taking the place of the water molecules associated with its surface and thus preserving its conformation.

Is trehalose the answer?
In 1971, John Crowe published a paper in the American Naturalist entitled 'Anhydrobiosis: an unsolved problem'. Some 20 years later,

and after extensive studies on the stabilisation of membranes by sugars and other compounds, he was led to declare that 'a single perturbation – synthesis of a disaccharide such as trehalose or sucrose – is sufficient to achieve survival'. In other words, the problem of anhydrobiosis had been solved and the answer was trehalose (or sucrose in plants). The evidence that trehalose stabilises membranes and proteins during desiccation is impressive. It is the most efficient compound of all those which have been tested in this respect. It must be said, however, that these studies have largely been conducted on isolated proteins, artificial membranes or on membranes which have been isolated from organisms which are clearly not anhydrobiotic (such as membranes from the muscles of lobsters). This is understandable, since it is difficult to demonstrate the activity of trehalose in an intact organism. However, the attachment of trehalose to membranes in an anhydrobiotic animal has yet to be demonstrated.

Trehalose is also produced by a number of animals which are not capable of anhydrobiosis. *Ascaris lumbricoides* is a large (10 centimetre) nematode which is parasitic in the intestine of humans. It contains trehalose within its body fluids and throughout many of its tissues, and yet is never exposed to desiccation. A number of other parasitic, as well as free-living, nematodes have been shown to produce trehalose. The main functions of trehalose in parasitic nematodes appear to be as a store of carbohydrates and as a blood sugar which is transported to provide the tissues with carbohydrates. Glucose is the sugar which is the main source of energy in organisms. Trehalose consists of two glucose units and can thus be used to store and transport glucose. The main carbohydrate store in animals is, however, glycogen. This is a much more efficient glucose store than trehalose, since it does not cause osmotic problems to the animal and it yields more energy than does trehalose. Parasitic nematodes produce both trehalose and glycogen, so what is the role of trehalose in this situation? Nematode parasites living in the intestine of their host are bathed in a solution of easily absorbed nutrients, including glucose. The trehalose may well assist in the uptake of glucose across the intestine of the nematode. Since trehalose does not cross cell membranes, the glucose can be locked up in the form of trehalose, which dissolves in the body fluids and can be transported to the tissues. This prevents glucose from leaking back across the wall of the intestine and being lost to the parasite. The nematode can thus maintain favourable conditions for the uptake of glucose from its host.

The ability of nematodes to synthesise trehalose may explain the widespread occurrence of anhydrobiosis in this group. Insects also produce trehalose, in which it is the major blood sugar. Despite the presence of trehalose, anhydrobiosis is rare in insects. Insects are certainly found in many situations where an ability to survive anhydrobiotically would be useful to them. If trehalose is the only adaptation needed, why is anhydrobiosis not more common in insects? Some recent studies on nematodes have cast further doubt on the proposition that trehalose is the sole adaptation necessary for anhydrobiotic survival. Some nematodes require slow drying even after they have completed the production of trehalose to concentrations found in the desiccated state. Other nematodes will not survive anhydrobiotically even though they produce trehalose in response to desiccation stress. It is clear that other, as yet unknown, adaptations are necessary for anhydrobiosis.

It is not to be denied that trehalose plays an important role in anhydrobiosis. It does not, however, appear to be the only adaptation which is necessary. Rather than considering the problem to have been solved, we need to explore what those other adaptations might be.

Recovering from anhydrobiosis
Surviving desiccation is only half the story. When water returns, the organism must absorb the water and resume active life. This could be as stressful as was the loss of water during desiccation, with the potential for material to be lost from cells and for disruption to occur. You might expect a dry nematode, for example, to take up water very quickly like a piece of dry cotton. Although the initial rate of water

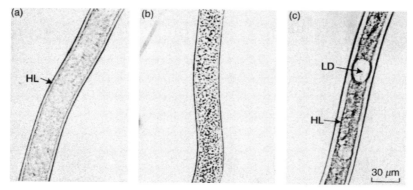

FIGURE 3.6 The appearance under the light microscope of the anhydrobiotic nematode *Ditylenchus dipsaci*. In (a), the nematode has not been exposed to desiccation; in (b) the nematode is completely desiccated (specimen mounted in a non-aqueous medium); and in (c) the nematode has recovered activity after a period of anhydrobiosis and then immersion in water for two hours. During rehydration, a hyaline layer (HL – comprising the muscle cells and epidermis) appears between the intestinal cells and the cuticle. Large globules within the intestinal cells also emerge, due to the fusion of smaller droplets of lipid (LD). From Wharton *et al.* (1985), reproduced with the permission of the *Journal of Zoology*.

uptake is indeed rapid, returning to a water content half that of normal within a few minutes, the rate of uptake slows and it takes several hours for the normal water content to be reached. Just as a slow rate of water loss was important to survive desiccation, a controlled process of recovery is necessary. In some cases, recovery is enhanced if the dry animals are exposed to moist air before immersion in water.

Anhydrobiotic nematodes do not start moving for several hours after they are reimmersed in water. This apparent period of inactivity is called 'the lag phase'. Although the nematodes do not move during the lag phase, they are certainly not inert. If you measure their metabolism, it commences immediately on immersion in water and rapidly rises to normal levels. During the lag phase, the nematode is metabolising like crazy but not moving – so what is it doing? In some nematodes, an orderly series of changes in their body structure has been observed. Most noticeable is the formation of large droplets in the intestinal cells and the appearance of a clear layer in between the intestine and the cuticle (Figure 3.6). The large droplets are due to small droplets of lipid fusing together to form larger ones, while the clear layer is due mainly to an increase in the thickness of the muscle cells of the animal. The structural changes are broadly the reverse of those observed during desiccation. The nematode appears to be undergoing some sort of process of repair, or restoration of a normal physiological state, during the lag phase that has to be completed before it can start moving again. If repair is occurring during the lag phase, you would expect its length to increase with more severe desiccation and hence with more damage to repair. This is indeed what happens, with the plant-parasitic nematode *Ditylenchus dipsaci* taking three times as long to recover after desiccation at 0 per cent relative humidity than after drying at 98 per cent relative humidity. It is the severity of the stress imposed during drying that determines the length of the lag phase, rather than the final relative humidity to which the nematode is exposed. A lag phase, the length of which appears to depend on the severity of desiccation, has also been observed in tardigrades.

There is more direct evidence for repair during the lag phase. The nematode cuticle acts a barrier to the exchange of materials with the outside world. This permeability barrier is partly destroyed by desiccation and then repaired during rehydration. The cells of some anhydrobiotes leak substances from within the cell when first immersed in water. The termination of this leakage could be due to the repair of membrane damage or to a physical change in membranes.

As well as repairing or restoring membranes and other structural elements, anhydrobiotic organisms will have to restore other aspects of the functioning of their cells to normal. The removal and then return of water will produce profound disturbances to the cell's internal environment. The water balance of the cell is obviously disrupted. After desiccation, the cell is shrunken. Sufficient water must be absorbed during rehydration to restore its normal water balance, but not too much or the cell will burst. The cell needs to restore its normal volume. All cells actively maintain a difference in the concentrations of various substances within the cell, compared with those outside the cell. As water is removed, the concentration of dissolved substances will increase and then decrease again when water returns. As well as restoring its water balance, the cell will also need to restore the normal concentrations of substances important to its functioning. This is of particular importance for muscle and nerve cells which depend on there being differences in the concentrations of particular ions (electrically charged atoms or molecules dissolved in water) between the inside and outside of the cell in order to function. Nerve impulses involve movements of sodium and potassium ions across the membranes of nerve cells, while muscle contraction depends on the movement of calcium ions. Restoring the concentration gradients of these substances is important for the nerves and muscles to work and this may explain the lack of movement by anhydrobiotes during the lag phase. Other features of the cell that may require restoration after anhydrobiosis include: pH, osmotic balance and the concentrations of oxygen and other dissolved gases.

The processes of repair and restoration may need to proceed to a certain level before the activity of anhydrobiotic animals will resume. Further repair may be necessary, however, before they will survive a further episode of desiccation. For some anhydrobiotes, survival has been shown to decline with repeated cycles of desiccation and rehydration. This may have been simply a result of the organism not being given sufficient time to recover before exposure to another bout of desiccation. The ability of animals to survive anhydrobiotically involves a suite of mechanisms, or potential mechanisms. These are summarised in Figure 3.7.

THE MECHANISMS OF ANHYDROBIOSIS IN PLANTS
Although the seeds, spores and pollen of many plants are capable of anhydrobiosis, it is difficult to distinguish the mechanisms which enable them to survive anhydrobiotically from those associated with their formation, germination and development. The tissues of resurrection plants and of desiccation-tolerant mosses, algae and lichens are, however, mature and provide systems in which the mechanisms of anhydrobiosis can be investigated (Figure 3.8). We would like many

NORMAL LIFE

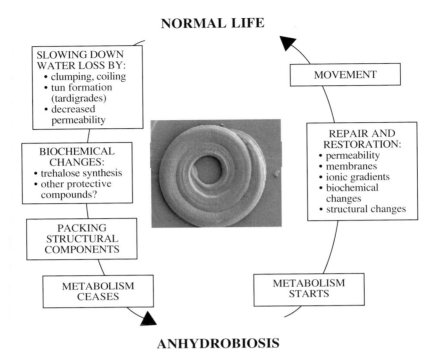

SLOWING DOWN
WATER LOSS BY:
• clumping, coiling
• tun formation
  (tardigrades)
• decreased
  permeability

MOVEMENT

BIOCHEMICAL
CHANGES:
• trehalose synthesis
• other protective
  compounds?

REPAIR AND
RESTORATION:
• permeability
• membranes
• ionic gradients
• biochemical
  changes
• structural changes

PACKING
STRUCTURAL
COMPONENTS

METABOLISM
CEASES

METABOLISM
STARTS

ANHYDROBIOSIS

FIGURE 3.7 The changes that occur during desiccation and rehydration of an anhydrobiotic animal, such as a nematode.

agriculturally important plants to be more drought tolerant. If the mechanisms of survival in anhydrobiotic plants can be understood, it may be possible to transfer these abilities to crop plants, using the techniques of genetic engineering, to allow them to be grown in more drought-prone areas.

Anhydrobiotic plants can be divided into two groups on the basis of the rate of desiccation which they can survive. The ability to survive fast rates of water loss is restricted to some species within the less complex groups of plants: mosses, liverworts, algae and lichens. These have little ability to conserve water and their water content rapidly comes into balance with that of their environment. Desiccation-tolerant higher plants (resurrection plants) require slow drying in order to survive anhydrobiotically and they have morphological and physio-

FIGURE 3.8 Changes in shoots of the African resurrection plant, *Myrothamnus flabellifola*, during rehydration. The branch on the left was air dry, the remainder (from left to right) were following four, eight, 20 and 24 hours of immersion of the stem in water. The leaves expand and rotate downwards as they rehydrate. The leaves of the fully hydrated plant are up to 8–10 millimetres long. Photo: Don Gaff.

logical mechanisms which ensure the slow rate of water loss necessary for their survival. These two groups of anhydrobiotic plants have been referred to as fully desiccation-tolerant plants and modified desiccation-tolerant plants respectively. Fully desiccation-tolerant plants must always be prepared for periods of anhydrobiosis, while the mechanisms which enable modified desiccation-tolerant plants to survive are mobilised only when they are exposed to desiccation. For the higher plants, the costs involved in diverting resources from growth and reproduction to survival are thus only incurred when they are faced with a life-threatening exposure to desiccation.

Structural preservation in the dry state varies in different species of anhydrobiotic plants. In some, their structure is well preserved and they retain chlorophyll, while in others there is extensive degradation of the structures within their cells and chlorophyll is lost. During rehydration, the normal cell structure takes several days to be restored and chlorophyll is resynthesised. There are thus two different strategies for surviving desiccation. Some species prevent damage during desiccation, while others repair the damage when water returns. For many species, some mixture of these two strategies is likely to be important.

A number of different mechanisms have been suggested to be important in the anhydrobiotic survival of resurrection plants. Trehalose is rare in plants, although its presence has been reported in a few species, but other sugars (such as sucrose, raffinose and maltose), polyols (such as glycerol, sorbitol and mannitol) or amino acids (such as proline) may play a similar role to trehalose in plants, protecting membranes and proteins during desiccation. The sugars may also be involved in the formation of glasses which stabilise the cells during anhydrobiosis. Glass formation has been demonstrated in plant seeds.

Biological reactions with oxygen can generate highly reactive charged molecules, particularly the superoxide radical (O_2^-) and the hydroxyl radical (OH^-). These radicals react destructively with membranes and biological molecules. Organisms have various mechanisms for protecting themselves from, and preventing the accumulation of, these reactive radicals. In particular, they have enzymes, such as superoxide dismutase, catalase and peroxidase, which facilitate their removal. These protective reactions may be lost during desiccation, since metabolism ceases, but resurrection plants may have an enhanced ability for the functioning of these enzymes to continue during water loss and rehydration. The loss of chlorophyll during the desiccation of some anhydrobiotic plants is interesting in this respect since photosynthesis involves reactions with oxygen which can generate these reactive radicals.

One approach to determining the mechanisms which might be important in stress resistance is to compare the pattern of protein production in the stressed and the unstressed organism. This approach, called proteomics, aims to look at the total pattern of proteins produced (the proteome). The proteome differs from the genome (the complement of genes) since not all proteins encoded by genes are produced at any one time. Any shift in the pattern of protein production, or gene expression, in response to a stress such as desiccation may indicate the mechanisms that are important in the response to that stress. This approach has yet to be applied to anhydrobiosis in animals, but there have been some promising results with the response of plants to drought and desiccation, including that of seeds and of resurrection plants.

Although the overall rate of protein synthesis eventually declines during exposure to drought, there are proteins synthesised which are either not produced by the plants when not under stress or which are produced in greater quantities. One group of these desiccation-induced proteins has been called 'dehydrins'. Dehydrins have been identified in seeds, in resurrection plants and in desiccation-tolerant mosses and liverworts. They have even been reported from several species of cyanobacteria and are thus found in a wide range of photosynthetic organisms. They are present in the nucleus, the chloroplasts and the cytoplasm of cells. In addition to plant tissues which are anhydrobiotic, dehydrins are found in a variety of plants during drought which, although drought resistant, will not survive anhydrobiotically. Many of the proteins produced by mature plants during drought are also produced by seeds in the later stages of their development (late embryogenesis-abundant proteins). Seeds dehydrate during the final stages of their development. The production of related proteins by seeds and by mature tissues during drought suggests that similar mechanisms are involved in surviving desiccation in both of these situations.

The synthesis of dehydrins by higher plants is triggered by desiccation and by exposure to abscisic acid, a plant hormone which is associated with plant responses to a variety of environmental stresses. Some of the changes observed in resurrection plants are, however, independent of the action of abscisic acid. The responses of plants to drought, cold and osmotic stress are similar since all these stresses involve problems of water availability. Exposure to low temperature induces a physiological drought since the increased resistance of roots to water movement means that the water lost by transpiration through the leaves exceeds that taken up by the roots. Many plants have been shown to synthesise dehydrins in response to cold stress. In mosses, which will survive rapid desiccation, dehydrins and sugars are present in fully hydrated tissue and protein synthesis is not triggered by desiccation. Thus, they have the compounds which protect them during desiccation permanently present in their tissues.

The function of many of these desiccation-induced proteins has yet to be deciphered. Some are proteases which may break down redundant proteins, thus releasing a source of amino acids which can be used for making new proteins. Others are involved in the synthetic pathways responsible for producing sugars and other compounds which are thought to act as protective substances during desiccation. Membrane proteins responsible for the transport of water and ions are likely to be important in controlling water flux during desiccation and rehydration. Some proteins may be involved in structural rearrangements, repair after rehydration and the removal of toxins (such as superoxide and hydroxyl radicals). Proteins that are also found in the final stages of seed formation are good at absorbing and holding on to water. They may thus enable the seed to maintain a minimum water content in the face of desiccation. They may also be involved in the protection of other proteins and, together with sugars, the formation of a glassy state which stabilises the structure of the tissue. Dehydrins have been found which are associated with membranes, suggesting a role in their stabilisation. They may interact with the surface of macromolecules and membranes, together with sugars, preserving their structural integrity during desiccation.

As well as mechanisms that protect cells during dehydration and in the dried state, repair following rehydration also appears to be important. Some anhydrobiotic plants show extensive structural disruption which is repaired on rehydration. Even where structural disruption is not visible, the leakage of cell contents into the surrounding water indicates that some membrane damage has occurred. In some resurrection plants, the production of several specific proteins associated with the rehydration phase has been demonstrated. In desiccation-tolerant mosses, there is no protein synthesis during desiccation but there is during rehydration. The metabolism of mosses rapidly recovers after rehydration, but the time taken to recover depends on the rate of water loss during desiccation. The more severe the desiccation stress, the longer it takes the moss to recover. This parallels observations on the lag phase of anhydrobiotic nematodes. The proteins which are synthesised during rehydration have been called 'rehydrins'. Their functions are unknown but they are thought to be involved in the process of repair of the damage that results from desiccation.

There are thus a variety of mechanisms that appear to be involved in anhydrobiosis in plants. Plants vary in their relative reliance on protective mechanisms during desiccation and repair mechanisms during rehydration. Protection involves both low molecular weight substances, such as sucrose, and proteins like dehydrins. In higher plants (resurrection plants), protective mechanisms are induced by desiccation stress, while in lower plants (mosses, liverworts, algae and lichens) the protective mechanisms are perpetually present. Although the mechanisms of anhydrobiosis and drought tolerance in plants are still not fully understood, it is clear that the application of the techniques of molecular biology to the problem is making good progress towards that goal. Such an approach has yet to be applied to the study of anhydrobiosis in animals, but, no doubt, some surprises await us when it is.

ANHYDROBIOSIS IN MICROORGANISMS

Bacteria are exposed to desiccation if they are carried into the air, are living on the surface of (or within) the soil when it dries, and if they are on or inside rocks and on the surface of plants and the skins of animals. Most bacteria have some degree of tolerance to desiccation. Some will survive after rapid drying, while others need a slow rate of water loss to survive. In general, they survive for longer after slow rather than fast drying. Spores, cysts and other sorts of resting stages survive better than do normal vegetative cells. These resting stages have some sort of structure outside the cell that forms a wall which protects it against environmental hazards, including desiccation. Viable bacteria have been isolated from sediments between 10 000 and 13 000 years old from ice cores at Vostok station in Antarctica. Under these conditions, the bacteria had been freeze dried for this period of time. Only spore-forming bacteria were found in the deeper, and therefore older, sediments.

As with animals, anhydrobiotic microorganisms accumulate trehalose. This is thought to stabilise membranes and proteins by replacing the water which plays a part in maintaining their structure and/or by forming a glass-like state which immobilises the cell's internal organisation. Other mechanisms may be involved in anhydrobiosis, such as preventing the accumulation of reactive oxygen radicals, repair mechanisms during rehydration and proteins which protect and assist in the folding of other proteins (molecular chaperones). Microorganisms may have less ability to synthesise specific proteins involved in the response to desiccation (like the dehydrins in plants), since (being single cells) their dehydration is likely to be much more rapid than that of a plant or animal. Many bacteria secrete large amounts of material to the outside of their cells (polysaccharides – large molecules consisting of repeating sugar subunits), which form a matrix or sheath surrounding them. These extracellular polysaccharides are hygroscopic (they absorb water). Thus, they may reduce the rate of water loss during desiccation and assist with water uptake during rehydration. They may also have protective properties, such as the formation of a glass, or some other stable state, during desiccation.

ANHYDROBIOSIS AND HUMANS

Although the idea of putting humans into suspended animation is likely to remain in the realms of science fiction, at least for some time to come, anhydrobiotic organisms affect human welfare and research on anhydrobiosis has produced, or has the potential to produce, technologies of use to us. The ability of many microorganisms to survive anhydrobiotically means they can be dispersed through the air. This results in the contamination and spoilage of food and in the spread of disease. Allergies (hayfever) result from the inhalation of airborne pollen and algae. The seeds of many weed plants are also dispersed by air. Plant-parasitic nematodes cause crop failure and reduced yields.

Their anhydrobiotic stages enable them to be dispersed in dry soil, seeds and plant debris and their resistance limits our ability to control these pests. The infective larvae of many animal parasites are also capable of anhydrobiosis.

On the positive side, the natural abilities of microorganisms and seeds to survive anhydrobiosis enable us to store them for long periods of time. The discovery of the properties of trehalose in stabilising membranes and proteins has led to its use in preserving a wide range of biological products. It has been successfully used to air dry and store antibodies, enzymes and blood coagulation factors. It may also be useful for preserving pharmaceutical products, vaccines and the systems which deliver these to their target sites; liposomes, artificial membrane spheres which are used as drug delivery systems, can be dried and stored with the aid of trehalose. The taste and properties of foods change upon air drying. If dried with trehalose, however, blended fresh eggs, fruit purées, herbs and even fruit slices retain more of the properties of the fresh product than if they are dried without trehalose. Attempts have been made to use trehalose to induce anhydrobiosis in cells, organs and even whole organisms which do not normally have this ability. Simply adding trehalose has had little success since it does not easily pass through cell membranes. The trehalose needs to be inside the cells to have any protective effect and, to preserve a membrane, it needs to attach to both of its sides. This could be achieved by providing cells with the ability to synthesise trehalose using the techniques of genetic engineering. Tobacco plants which have had the enzyme trehalose synthase from yeast inserted into their cells show an increased resistance to drought, by the development of drought avoidance mechanisms.

Drawing on tardigrades for their inspiration, Kunihiro Seki and his colleagues at Kanagawa University in Japan have flushed trehalose through rat hearts before desiccating them by packing them in silica gel which absorbs water. They then perfused the hearts with perfluorocarbon, a biologically inert compound, and stored them at 4 °C in airtight jars. After 10 days, the team were able to revive the hearts and get them beating again. The reasons for their success are a bit perplexing, since the trehalose is unlikely to have penetrated the cells of the heart. However, if the integrity of hearts, and other organs, can be maintained during long-term storage, there are obvious implications for transplant surgery.

We would like to be able to improve our methods of storing biological materials for a variety of medical, commercial and conservation applications. If we can understand how anhydrobiotic organisms survive in their remarkable way, we can apply this knowledge to the storage of cells, organs and organisms which do not naturally possess this ability.

HAS THE PROBLEM OF ANHYDROBIOSIS BEEN SOLVED?

Although we have some important clues as to how organisms survive anhydrobiosis, it is clear that we do not, as yet, fully understand the phenomenon. The problem is indeed intimidating, as Michael Potts of the Virginia Polytechnic Institute and State University was led to comment with respect to bacteria: 'the prospects for explaining how a single cell with a complement of some 3000 proteins can have the bulk of its water removed, remain desiccated for perhaps tens of years . . . and then resume coordinated metabolic activities within seconds of rehydration is daunting at best!' Explaining the survival of the much more complex plants and animals is even more difficult.

There is clearly an association between trehalose or sucrose and anhydrobiosis. However, there are many examples where trehalose or sucrose are produced and yet the organism does not survive anhydrobiosis. While trehalose or sucrose appears to be necessary for anhydrobiosis, it does not appear to be in itself sufficient. The challenge is now to understand what other mechanisms are involved. Although trehalose can stabilise proteins and membranes in the desiccated cell, we need to investigate how anhydrobiotic organisms solve the higher order problems of metabolic and physiological integration during desiccation and rehydration. There are many problems to be solved before we fully understand the phenomenon of life without water.

4 The hot club

Changes in temperature have profound effects on biological processes. In general, the rate at which life processes (such as metabolism, respiration and growth) proceed declines as the temperature decreases and elevates as it increases. This is because high temperatures supply more kinetic energy to reactions than do low temperatures, allowing the reacting molecules to come into contact and to interact more frequently. If the temperature continues to increase, however, it starts to have destructive effects on organisms. The rate of metabolism, and other life processes, declines again. There is thus an optimum temperature at which metabolism is at its greatest rate and, if the temperature decreases or increases from this optimum, the metabolic rate declines (see Figure 1.1). As was pointed out in Chapter 1, the decline in metabolic rate at low temperatures (below the optimum) is due to very different causes from those producing a decline at high temperatures (above the optimum). The rate declines at low temperatures mainly because there is less kinetic energy driving the metabolic reactions, but the effect is largely reversible (unless freezing or some other event occurs which the organism cannot survive; see Chapter 5). The decline in metabolic rate at high temperatures is, however, due to destructive effects on the proteins, and the other components, of the organism. Some of these destructive effects may be reversible if the temperatures are not too high. At higher temperatures, irreversible changes occur or there is too much damage to repair and the organism will die.

Central heating and solar heating
The vast majority of organisms have little ability to control the temperature of their bodies and their temperature is the same as that of

their immediate surroundings. Most of their heat is absorbed from outside their bodies (they are ectotherms). In most situations, their ultimate source of heat is solar radiation. Other external heat sources include biological combustion processes (such as the heat generated within a compost heap) and geothermal activity (in hot springs, hydro-thermal vents and deep underground). Endotherms have their own central heating system. They generate heat via their metabolism and are able to maintain a body temperature which is above that of their surroundings. They will also, of course, absorb heat from the sun and from their environment and thus have both external and internal sources of heat. The metabolism of ectotherms will also generate some heat, but their metabolic rate is not always sufficient to raise their body temperature above that of their surroundings and the heat soon dissi-pates. Colouration and orientation to the sun affects the rate at which organisms gain or lose heat. They may also have insulating structures which reduce heat loss (hair, feathers or scales) or structures which shield them from the sun and reflect its rays – thus reducing heat gain (reflective spines and surfaces).

Some ectotherms have behaviours or growth forms which provide a limited degree of control over their internal temperature. They can do this simply by moving from a hot place to a cooler place (or vice versa) by seeking out shade, burrowing into the ground or by emerging into the sun. Desert lizards bask in the sun or press their bodies against the warm sand to raise their body temperature in the morning. In the heat of the day, they stand erect, adopting a posture which increases heat loss from their body, or they simply retreat into the shade. These behavioural mechanisms give lizards a remarkable degree of control over their internal temperature (for an ectotherm).

Birds and mammals are the only organisms which can maintain a constant internal temperature in the face of changing external temper-atures. They are able to maintain a temperature higher than that of their surroundings since they generate heat internally by burning food through their metabolism. This is expensive in terms of energy or food use. A human at rest will consume 1300–1800 kilocalories per day at

20 °C, about 90 per cent of which is used to maintain body temperature. An endotherm will generally use about 10 times more energy at rest than does an ectotherm at the same temperature. The energy demands of metabolic heat generation are kept to a reasonable level, however, since the efficient insulation provided by fur or feathers and a layer of fat beneath the skin, together with other heat conservation measures, reduces the rate of heat loss from the body. Maintenance of internal temperature by endotherms does not, therefore, consume all their food reserves. The energy demands of maintaining a high internal temperature mean that there is a lower size limit on animals that can sustain this. A small animal loses heat faster than a large animal because of its greater surface area relative to its volume. A small animal therefore has to have a higher metabolic rate and consume more food to maintain its temperature than does a large animal. Shrews are the smallest mammals and hummingbirds are the smallest birds. These small endotherms have to eat constantly in order to satisfy the energy demands involved in maintaining their temperature.

Mammals and birds usually maintain a core temperature of around 37 °C and 41 °C, respectively. This is a relatively high temperature in the context of the range of temperatures found on Earth and, in most situations, the internal temperature of a bird or mammal is higher than that of its surroundings. Why do they maintain such a high temperature? One reason might be that, since their temperature is higher than that of their environment, they can lose heat by radiation, conduction or convection to regulate their internal temperature. The higher internal temperature also allows enzymes to catalyse biological reactions at a faster rate. This results in faster rates of growth and development. The internal temperatures of birds and mammals are about as high as they can be without incurring some of the harmful effects of high temperatures. Maintaining a constant temperature may have allowed enzymes, and biological processes, to have become optimised to function best at the body's internal temperature. A high internal temperature also enables these animals to function efficiently even though the external temperature is low. This allows activity at night and in cold

HEAT GAIN   HEAT LOSS

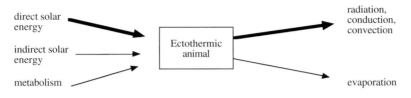

Too hot: seek shelter

Too cold: bask in the sun or press against hot surfaces

HEAT GAIN   **HEAT LOSS**

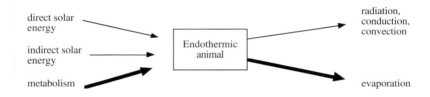

Too hot: seek shelter, decrease metabolic rate, divert
blood supply to surface, increase evaporation
e.g. by sweating, panting

Too cold: increase metabolic rate, retain heat through
insulation, divert blood supply away from surface

FIGURE 4.1 Comparing the mechanisms of heat gain and loss in an ectothermic
animal, such as a lizard, and an endothermic animal, such as an Emperor penguin or
a camel. For an ectotherm, solar radiation, either direct or via its heating of rocks or
soil, is the most important source of heat while, for an endotherm, heat generated
metabolically is more important. An ectotherm loses heat mainly by the processes
of conduction, convection or radiation from its body whereas, for an endotherm,
evaporation (such as sweating or panting) is the main method. Ectotherms rely
mainly on behavioural mechanisms to regulate their temperature while
endotherms use mainly physiological mechanisms.

environments. Endotherms are thus able to colonise habitats and environmental niches which were not open to ectotherms.

Some fishes, reptiles, invertebrates and even plants can be endothermic, at least for a period of time and in some parts of their bodies. Some large moths, for example, can generate heat internally which enables them to warm to a temperature at which they can fly even on a cold day or at night. Honeybees, as well as generating heat, huddle together to conserve heat and can maintain a fairly constant temperature within their hive by changing the density of the huddling. The heat is circulated and distributed within the hive by individuals moving from the warm interior to its cooler outer edges.

Animals have a variety of mechanisms for regulating their temperature and/or controlling the exchange of heat with their environment (Figure 4.1). The mechanisms they use depend on whether they are capable of generating their own heat and maintaining their temperature above that of their surroundings (endothermic) or are dependent on absorbing heat from the environment (ectothermic), and whether the conditions in their environment mean they need to gain heat (or prevent it from being lost) or lose heat (or prevent it from being gained).

Dying from the heat

High temperatures have destructive effects on the molecules which make up the cells of an organism. The properties of proteins depend on their three-dimensional shape (their folding or conformation). This shape is maintained by weak bonds between parts of their structure and by the interaction of the protein molecule with water molecules. These weak bonds and interactions require only a small amount of thermal energy to disrupt them. In fact, individual weak bonds are being broken and reformed continuously, even at normal temperatures. High temperature results in the bonds being broken at a greater rate than they are reformed. The three-dimensional shape of the protein thus breaks down, and it unfolds and can no longer fulfil its biological functions. The life-supporting processes of the organisms are thus not sustained and this leads to the death of cells and the death of the organism. The breakdown in the structure of a protein is called denaturation. It is seen when we poach an egg. The white of the egg contains the protein albumin. Before cooking, it is clear and fluid. As it cooks, it gradually becomes hard and opaque due to the denaturation (unfolding) of the protein by the heat. I wonder why we prefer to eat proteins denatured (cooked)? Perhaps humans learnt by experience that cooking destroyed the disease-causing organisms that may be present in food and so cooked food came to taste better to us. Certainly, parasitologists always like their steaks well done. Or perhaps cooked food is just easier to digest.

Some proteins are more heat stable than others and the process of denaturation can be reversible. As the protein cools, the weak bonds reform and its folding and functionality are restored (renaturation). The ability of proteins to recover their activity after heating will depend on the extent to which they have been denatured (the degree of unfolding). This will increase with temperature and with the time of exposure to high temperatures. If too much damage occurs, the proteins cannot renature and recover their biological properties sufficiently quickly to prevent the death of the organism.

It may, however, be difficult to relate the destructive effects of heat at the molecular level to the death of the whole organism, particularly for multicellular plants and animals. Animals or plants may die at temperatures below that at which their enzymes become denatured, as a result of heat-induced problems at a higher level of organisation than the molecular or cellular level. For terrestrial plants and animals, it may not be the high temperatures themselves which kill them, but rather the resulting loss of water. High temperatures may also result in a lethal imbalance between the various components of the organism's metabolism. Plants, for example, respire faster than they photosynthesise at high temperatures. They thus consume food at a greater rate than they can produce it and starve to death. A number of studies on fish indicate that a breakdown in their osmoregulatory systems (control of water and salt balance) is an important early event in heat stress. The nervous system seems to be particularly susceptible to disturbance by high temperatures and a loss of nervous integration and control would be fatal for an animal. It is not easy to define the temperature at which an organism dies since the lethal effects depend on how long it has been exposed to that temperature and on its thermal history – what it has experienced before exposure to a potentially lethal temperature.

Keeping cool

Faced with rising temperatures, some organisms will try to stay cool and maintain their internal temperature at a non-lethal level by losing heat from their bodies. Heat is lost from the surface of an animal and so an animal can increase its heat loss by exposing more of its surface to the air and by diverting more blood flow to beneath the skin of exposed areas. A more upright posture will expose a greater surface area and raising feathers or hair will allow more air circulation close to the skin. An elephant loses heat by flapping its ears, which are well supplied with blood vessels. Many desert animals have large ears, such as the fennec fox of the Sahara Desert, and can lose heat by diverting blood flow to these structures. Mammals which live in hot climates generally have a greater proportion of their surface exposed to the air (free of hair) than do those in cold climates.

The evaporation of water requires the absorption of heat. Terrestrial organisms can thus achieve cooling by evaporating water from their surface or from respiratory organs. Panting increases airflow over the tongue and through the mouth and respiratory system. This raises the rate of evaporation from these surfaces and produces cooling. The incoming air is at a lower temperature than the expired air. Warming and humidifying the inspired air absorbs heat from the body. Dogs use panting as their main method of unloading heat. They pant at a rate of 300–400 times a minute when hot, compared with 10–40 times a minute under cold conditions. Throat fluttering by birds has a similar cooling effect. Some small mammals cool themselves by spreading saliva over their fur.

Sweating produces evaporative cooling from the water secreted onto the surface of the skin. This is well developed in cattle, horses, hippopotamuses and humans, who sweat over their entire bodies. Sweating is poorly developed, or absent altogether, in other mammals (such as rodents and carnivores). There is a relationship between the relative importance of panting and sweating in mammals. Those which sweat over their whole bodies do not pant (e.g. humans) while those in which sweating is poorly developed have a high capacity for panting (e.g. dogs).

All these methods of heat loss also result in greatly increased rates of water loss. Sustained exposure to high temperatures is likely to result in death through dehydration, unless the animal has access to adequate supplies of water. Water conservation measures are essential for animals in hot dry environments. Small desert mammals do not usually unload heat by sweating or panting, since they could not survive the resulting water loss. Instead, they avoid the heat of the day in burrows or rock crevices and are active at night.

Marine animals have fewer options for controlling their temperature than do terrestrial animals. Sweating and panting will not produce cooling since water cannot evaporate from their surfaces. However, the marine environment produces a much more stable temperature environment, with the bulk and high heat capacity of the water providing a buffer against temperature change. Marine organisms are thus, in general, unlikely to experience extreme high temperatures or rapid changes in temperature, except around hydrothermal vents.

Getting used to the heat

If an organism is exposed to an elevated, but not fatal, temperature for a short period of time, there is a dramatic increase in its ability to survive high temperatures which would be lethal without this pretreatment. This is known as induced thermotolerance, or the heat shock response, and has been observed in almost every organism studied, from bacteria to mammals. The heat pretreatment produces, within minutes, a burst of protein synthesis. The proteins that are produced are absent, or are only present in low concentrations, in the unstressed state and are known collectively as heat shock proteins. These protect organisms against the harmful effects of heat, particularly by their effects on other proteins. Some heat shock proteins remove heat-damaged proteins from cells, preventing them from accumulating and interfering with the operation of the cell. Others act as molecular chaperones which help proteins maintain their conformation and resist denaturation by heat. Heat shock proteins also help proteins to refold and regain their function after a heat stress. Molecular chaperones are also present at normal temperatures where they assist protein folding and prevent proteins, which are present at high concentrations within cells, from sticking together and interfering with each others' formation and functions. The problems of protein folding and aggregation become worse at high temperatures and elevated levels of molecular chaperones are produced to deal with this as part of the heat shock response.

The heat shock response is an emergency system which provides organisms with a measure of protection against unpredictable stresses. Some environmental changes, such as a seasonal change in temperature, occur more gradually and with a greater degree of predictability. Over a period of days or weeks, the organism undergoes physiological changes which alter its tolerance to environmental stresses in a way which matches the changing season. The bullhead catfish, for example, can tolerate temperatures up to 36 °C during summer whereas, in winter, temperatures above 28 °C are lethal. This process of a change in tolerance is referred to as 'acclimatisation' if it occurs in nature and 'acclimation' if it is induced artificially in the laboratory. The acclimatisation process is complex and may involve a variety of changes. Some organisms produce different variants of enzymes which have the same function but different temperature optima. By shifting from one variant to another, the organism can adjust its operation to match the changing season. The properties of membranes may change by altering the proportions of different lipids in their structure. The seasonal production of some low molecular weight compounds (such as glycerol, proline and trehalose) may protect against environmental stresses. Temperature acclimatisation may even involve extensive reordering of the composition of cells. In some fish, for example, the proportion of the cell's volume that is occupied by mitochondria increases in winter and decreases in summer.

THE MEMBERS OF THE HOT CLUB

The ability of birds and mammals to regulate their internal temperature allows them to survive in some hot environments. Other organisms can do so by behavioural or structural mechanisms which allow them to lose, or avoid gaining, heat. Some organisms, however, lack these mechanisms and their lifestyle or habitat means that they have had to join the 'hot club' and survive high internal temperatures.

What's hot and what's not?

The great majority of organisms live in environments where temperatures are in the range of 0–48 °C. There are few habitats where temperatures rise above 48 °C and fewer still where temperatures are constantly above 48 °C. Temperatures below 0 °C are more common, exposing organisms to freezing or the threat of freezing. How organisms cope with temperatures below 0 °C will be described in the next chapter. Few animals or plants can survive sustained exposure to temperatures above 48 °C for more than a short period. Such an ability might thus be a useful criterion for membership of the hot club. Birds maintain an internal temperature of 41 °C, only seven degrees below the cut-off point. Since they happily survive this temperature inside their bodies, the destructive effects must kick in rather soon as temperatures rise. The lethal temperature is only a few degrees above that which is optimum for their metabolism. Some proteins start to be denatured at 45 °C and most animals die after exposure to 50 °C for more than brief periods.

In hot subtropical deserts, air temperatures regularly reach 50 °C and ground temperatures 70 °C, absorbing the baking heat of the sun. As we saw in Chapter 2, however, desert plants and animals either avoid the heat or have mechanisms which prevent them from heating up to lethal levels. Desert microorganisms which cannot avoid the heat or cool themselves down may have to survive exposure to these high temperatures. Along with the heat, however, is likely to come desiccation. Anhydrobiotic organisms in the dry state can resist much higher temperatures than when they are hydrated (see Chapter 3). Protein denaturation by heat involves the molecule unravelling as the increased kinetic energy breaks the weak bonds which maintain its shape. This cannot happen in the absence of water, and anhydrobiosis also confers heat stability. Desert microorganisms are thus likely to survive high temperatures in a state of anhydrobiosis. Some anhydrobiotic animals (nematodes, rotifers, tardigrades and some arthropod larvae) and plants (resurrection plants and seeds) are also able to survive high desert temperatures in a desiccated state. These organisms use resistance adaptation – lying dormant until water, and lower temperatures, return.

Terrestrial hot springs and deep-sea hydrothermal vents are aquatic habitats and the organisms which inhabit them cannot rely on anhydrobiosis to survive (since they cannot desiccate). The conditions are also fairly constantly hot and the organisms use capacity adaptation to adapt to, and thrive in, the hot environment. They are thus extremophiles (thermophiles). There are no plants or animals that can survive these conditions which are constantly above 50 °C (or maybe just one animal, the Pompeii worm – see later). Membership of this hot club is restricted to microorganisms.

Hot club plants and animals

Desert animals which cannot avoid the heat and which do not have mechanisms to cool themselves down may reach temperatures (above 48 °C) that qualify them for membership of the hot club. The small desert snail *Sphicterochila boisseri* inhabits the Negev Desert of Israel. In dry conditions, it becomes dormant on the surface of sand and rocks where it is exposed to the full heat of the sun. The surroundings of the snail reach temperatures as high as 65 °C on the ground and 43 °C in the air. The shell is very shiny and reflects much of the solar radiation and the snail retreats to the upper whorls of the shell, leaving an air pocket

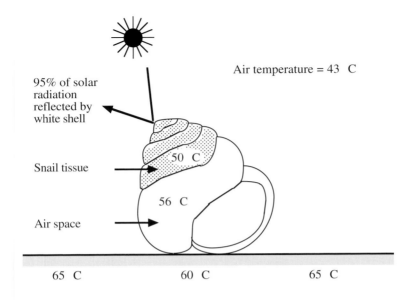

FIGURE 4.2 Temperature relations of the desert snail *Sphicterochila boisseri*. Redrawn from Schmidt-Nielsen *et al.*, 1971 and reproduced with the permission of the Company of Biologists Ltd.

which insulates it against heat gain from the ground. The snail's tissues, nevertheless, reach temperatures as high as 50°C, but it can survive this since its lethal temperature is between 50°C and 55°C (Figure 4.2). The animal is, however, dormant under these conditions and it is using a resistance adaptation. The snail does not desiccate appreciably and can maintain a water content of 76 per cent for several weeks under hot dry conditions. A desert ant, *Ocymyrmex barbiger*, forages during the day when temperatures on the surface of the sand are as high as 67 °C. The ant can avoid the extreme heat by raising its thin body above the surface on its legs and by seeking out shade and cooler areas. In spite of this, it survives body temperatures in excess of 52 °C for short periods, although this is close to its thermal death point. Foraging during the day is risky for the ant and many of them die from heat stress, but it allows them to be first at the scene to scavenge on the carcasses of other insects that have died from the heat. They also avoid the attentions of lizards and other predators which cannot sustain activity during the hottest part of the day.

Desert plants cannot move away and avoid exposure to the sun and heat. *Tidestromia oblongifolia* is a herbaceous perennial which grows during the summer in Death Valley, California, at temperatures in excess of 50 °C. It cannot grow in more moderate climates near the coast and is thus clearly heat adapted. Desert plants reduce heat gain by having a light colour and spines, hairs and waxy cuticles which reflect the sun's rays or shade the surface of the plant. Some plants (including *Tidestromia*) can cool down by having rapid rates of transpiration (which produces evaporative cooling at their surface). Desert succulents (including cacti), however, conserve their water and could not survive the water loss involved in evaporative cooling. Their internal tissues can reach temperatures above 60 °C and the plants can survive these temperatures for prolonged periods. One of the highest temperatures recorded in a live plant is 65 °C in a species of *Opuntia* (a cactus).

There are few animals or plants which can live in terrestrial geothermal springs. Brine flies will land and feed on the algae growing in the springs but only where temperatures have cooled to below 40 °C. The flies lay eggs that hatch and the larvae which emerge also feed on the algae. The larvae die, however, if the flow of water in the spring changes so that they are swept away or exposed to water which is hotter than that which they can tolerate. A nematode (*Aphelenchoides parientus*) has been reported from a hot spring living at temperatures of 45–51 °C and insect larvae (chironomids) from hot springs at 49–51 °C.

The water issuing from deep-sea hydrothermal vents can exit, superheated because of the high pressure at depth, at temperatures over 300 °C. It rapidly cools, however, since the surrounding seawater is at about 2 °C. The animals associated with the vents live mainly in waters at around 30 °C. Perhaps the best animal candidate for membership of the hot club is the Pompeii worm (*Alvinella pompejana*, a small polychaete annelid) which lives on the walls of black smokers, the chimney-like structures which form by the deposition of minerals around some deep-sea hydrothermal vents. The worm builds a tube, in which it lives, on the walls of the chimney. Hot water from the vent flushes through the tube, cooling as it mixes with the cold water at its mouth. Probes inserted into the tube using a 'deep submergence vehicle', by Craig Cary from the University of Delaware and his coworkers, have measured temperatures averaging 81 °C at the base of the tube and 22 °C at its mouth. When probes were positioned at the tail end of the worm, temperatures averaging 68 °C, and occasionally exceeding 81 °C, were measured. If these measurements represent the temperature of the worm itself, the Pompeii worm would not only be the most thermotolerant multicellular (and eukaryotic) organism known, but also one which can survive an unprecedented thermal gradient of up to 60 °C (Figure 4.3). Given the marked temperature gradient within the tube, however, the precise relationship between the position of the worm and that of the temperature probe is critical. The measurements were made using probes positioned by an operator in a deep submergence vehicle. The precise placing of the probe and the monitoring of its position in relation to the worm was therefore difficult. Craig Cary tells me that their most recent temperature measurements, however, confirm their earlier findings and that they have recorded temperature spikes up to 89 °C.

Another approach to determining the tolerance of the Pompeii worm is to examine the thermal stability of its enzymes and the other molecules and components which are vital for its survival. The worm's haemoglobin becomes unstable at temperatures above 50 °C and the respiration of isolated mitochondria ceases above 49 °C. Denaturation of its cuticle collagen (the protein which is the main structural component of the body wall) occurs around 45 °C. These measurements, and the temperature optima of its enzymes, suggest that the worm is limited to temperatures below 50 °C, with an optimum body temperature of 30–35 °C. Craig Cary has, however, isolated an enzyme from bacteria which are associated with the surface of the worm that has a temperature optimum of about 82 °C. It is difficult to relate these measurements on isolated components to the actual thermal

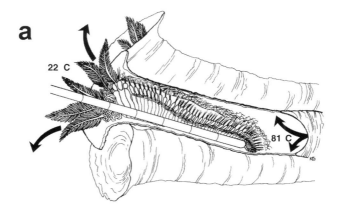

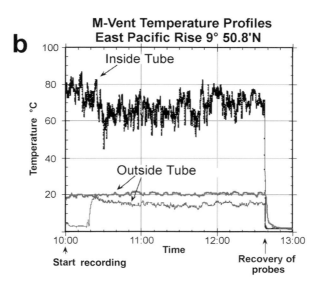

FIGURE 4.3 Temperature recordings from the tubes of the Pompeii worm, *Alvinella pompejana.* The worm is about 6 centimetres long.

(a) Drawing showing the worm in its tube with a temperature probe in position. The temperatures shown are the maximum recorded during one deployment of the probe.

(b) Temperature recordings from inside the tube and two recordings from outside the tube, taken over a three-hour period.

Reprinted by permission from Nature (Cary *et al.*, 1998) copyright (1998) Macmillan Magazines Ltd. Graphic kindly supplied by Craig Cary of the University of Delaware.

environment of the worm. In the intact animal, protective thermal mechanisms may operate (such as via molecular chaperones). If there is indeed a marked temperature gradient along the body of the worm, thermally sensitive functions may be centred in the cooler parts of the animal. There is clearly more to learn about this fascinating animal.

Hot club microbes

New Zealand, my adopted home, lies on the boundary of two of the great continental plates which make up the surface of the Earth. Here, the Pacific Plate is drawn under the edge of the Indo-Australian Plate, in a process known as subduction, creating a fault line which runs for more than 600 kilometres through both the South and North Islands. The forces generated by the subduction process have created huge mountains (the Southern Alps) and regular earthquakes. As the Pacific Plate is dragged beneath the Indo-Australian Plate, it is forced deeper into the Earth, encountering hot temperatures which melt its rocks and convert them into molten magma. In places, this finds its way to the surface via cracks in the crust to form volcanoes. The active volcanoes of New Zealand, White Island and Mounts Ruapehu and Ngauruhoe, are part of the Ring of Fire – a chain of volcanoes around the edge of the Pacific. When water comes into contact with the hot rocks, it may emerge at the surface in the form of bubbling mud pools, geysers and hot springs at temperatures up to 101 °C. The area around the lake-side town of Rotorua in the central North Island is New Zealand's best known thermal area. The whole town is pervaded with the bad eggs smell of hydrogen sulphide gas (known affectionately by the locals as 'Soir de Rotorua') and one area was named Whangapipiro ('an evil-smelling place') by the indigenous Maori people. Bubbling mud pools are one of the more unusual hazards to be encountered on the local golf courses. Hot springs are found throughout the world, associated with subduction zones and volcanic activity, with their greatest concentrations in the western USA (including Yellowstone National Park), Iceland, New Zealand, Japan, the Mediterranean countries, Indonesia, Central America and Central Africa.

There are no animals or plants in these hot springs, in which the water is too hot to immerse your hand, but these scaldingly hot waters are not devoid of life. The presence of bacteria is revealed by, often brightly coloured, films and stringy filaments. If glass slides are immersed in the water, they rapidly become colonised by bacteria. In the 1960s, Tom Brock, now retired from the University of Wisconsin–Madison, started isolating and culturing bacteria from the hot springs of Yellowstone National Park. He found many different types of thermophilic bacteria, including several new species which lived at temperatures in excess of 73 °C. Large yellow masses of filaments, from which a bacterium identified as *Thermus aquaticus* can be isolated, grow in the hot springs of Yellowstone at temperatures of 75–80 °C. Some of the hot springs of New Zealand reach higher temperatures than those of Yellowstone because of their lower altitude. On a visit to New Zealand, Tom Brock was able to isolate bacteria at temperatures of around 100 °C. He could not, however, isolate any organisms from fumaroles (crevices in the walls of a volcano through which steam issues) in Italy and Iceland which produced superheated steam at temperatures much greater than 100 °C. It appears that the bacteria need the presence of liquid water and that the boiling point of water is the limit for life in terrestrial hot springs.

The presence of bacteria in boiling water is limited to hot springs where the water is at an alkaline or neutral pH. Acidic conditions are more difficult to cope with and the upper temperature limit for organisms in acidic hot springs is about 90 °C. The high pressure in the deep sea allows water to emerge superheated from hydrothermal vents at temperatures above 300 °C. Bacteria growing near the vents have been isolated at temperatures of about 115 °C. These have been called hyperthermophiles, extreme thermophiles or ultrathermophiles – all ways of saying 'boy are they hot!'. Some of these, such as *Pyrodictium* and *Pyrolobus*, have optimum growth temperatures above 100 °C (105 °C). Hyperthermophilic bacteria have also been isolated from the production fluids of oil fields at temperatures up to 110 °C. Even hotter conditions, which nevertheless support bacterial growth, may be found in the deep subsurface. It is thought that the temperature limit for bacterial life may be up to 150 °C.

Only bacteria can survive extreme thermophilic conditions. The upper limit for the growth of thermophilic fungi is around 60 °C, the maximum growth temperature of the alga *Cyanidium caldarium* is 57 °C and protozoa have been grown at 56 °C. Thermophilic and hyperthermophilic bacteria are found in both the eubacteria and the archaea. Most hyperthermophilic archaea are associated with waters or soils which are heated by geothermal activity and which contain sulphides and elemental sulphur. Reactions involving sulphur are an important part of the metabolism of most of these organisms.

HOT AND LOVING IT

Thermophilic and hyperthermophilic eubacteria and archaea not only survive hot temperatures but need them for their growth and reproduction. The optimum growth temperature of thermophiles is above 40 °C and that of hyperthermophiles is between 80 °C and 100 °C. Hyperthermophiles generally do not grow below 60 °C. They are thus using capacity adaptation and their biological systems are adapted to operate at these high temperatures. How can these organisms live and survive at temperatures which are fatal to most others on Earth?

Thermostable molecules and membranes

Eubacteria and archaea are small and lack insulation. The inside of their cells are thus at the same temperature as that of their surroundings. Since many proteins denature and cease to function at temperatures above about 50 °C, the proteins involved in the structure and metabolism of thermophiles and hyperthermophiles must either be exceptionally stable at high temperatures or are resynthesised at the same rate as they are degraded. To resynthesise their components continually would be very expensive for a thermophile, in terms of using their resources, and so producing thermostable enzymes, and other proteins, is likely to be the favoured strategy. The other types of molecules, and structures such as membranes, that make up the cells of thermophiles also need to be thermostable.

The composition of the membranes of archaeal cells is fundamen-

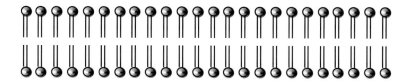

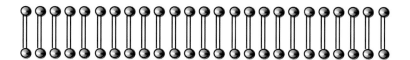

FIGURE 4.4 The lipids of the membranes of bacterial and eukaryotic cells (top) form a bilayer, while those of archaeal cells (bottom) form a monolayer. Only the lipid component of the membrane is shown.

tally different from that of bacterial and eukaryotic cells. In the latter two groups, their membranes consist of two layers of lipid, with associated proteins and carbohydrates. In the archaea, the lipid component of the membrane consists of a single layer (Figure 4.4). A lipid monolayer is less likely to split apart than is a bilayer and hence is more heat stable. This membrane stability contributes to the ability of hyperthermophilic archaea to survive higher temperatures than do thermophilic eubacteria or eukaryotes. As well as their membranes having a different structure, the lipids of archaeal membranes also have a different chemical composition from those of bacterial and eukaryotic cells. The lipids of thermophilic eubacteria are also different from those of bacteria which grow at more normal temperatures. Their membrane lipids are rich in saturated fatty acids (one of the components of lipids). Saturated fatty acids form stronger bonds between their molecules than do unsaturated fatty acids. The lipid composition of thermophiles thus makes their membranes more stable at high temperatures. However, in order to fulfil its function, a biological membrane must have a degree of fluidity. This allows control over the exchange of substances between the cell and its environment. The composition of the membranes of thermophiles means that their membranes have the required fluidity at high temperatures, without them disintegrating. However, if the temperature falls, the membrane solidifies, loses its fluidity and the cell can no longer exchange materials with its environment. Growth will then cease and the cell becomes dormant or dies.

The enzymes, and other proteins, of thermophiles are not only heat stable but they function best at high temperatures. This thermal stability and high optimum temperature is achieved by only a few changes in the amino acid sequence which makes up the structure of the protein. This causes the protein to fold in a way which increases its resistance to denaturation by heat. An increased number of bonds or bridges between the different parts of the protein molecule contributes to this stability, as does a more tightly packed interior which resists unfolding. An enzyme needs a degree of flexibility in order to interact with the molecules involved in the reaction which it catalyses. The increased thermal stability of the enzymes of thermophiles means that they have the required flexibility, and hence function best, at high temperatures. At lower temperatures, they become too inflexible to function and the growth of cells will cease.

Hyperthermophiles also need to prevent their genes from melting. The DNA of most organisms would melt at temperatures above 90 °C. The DNA of some hyperthermophiles is made more heat stable by adjusting the proportions of some of the components (bases) which make up its structure. The DNA molecule of some hyperthermophiles is supercoiled, a configuration in which the double-stranded DNA molecules are further twisted. This presumably increases its heat stability.

Repair and protection mechanisms
As well as having heat stable molecules and membranes, thermophiles produce substances which protect their cells or which repair heat-

induced damage. The production of some low molecular weight compounds, such as trehalose and 2,3-diphosphoglycerate, is associated with thermotolerance in some organisms. Molecular chaperones are produced by thermophiles, as they are by all organisms, and help to stabilise and refold proteins as they start to denature near the upper limit of the organism's temperature range. Proteins which bind to DNA (such as histones) may contribute to thermal stability.

Hot properties

The extreme thermal stability of the enzymes of hypertheromophiles has earned them the name 'extremozymes'. For example, ferredoxin from *Pyrococcus furiosus* is most active at 118 °C and does not denature until 140 °C. A wide variety of extremozymes have been isolated from various hyperthermophiles. These are of interest as biological catalysts for a variety of industrial applications involving biological processes which function best at high temperatures – from laundry and dry cleaning through to food processing and pharmaceutical manufacture.

One example of an extremozyme which is of great practical importance is the DNA polymerase isolated from *Thermus aquaticus* (known as Taq polymerase). This is used in the automation of the repetitive steps in the polymerase chain reaction (PCR) which multiplies or amplifies specific DNA sequences for applications such as DNA fingerprinting. The PCR technique involves several cycles of exposure to high temperatures to dissociate the DNA before allowing it to cool and repolymerise. The use of a heat-sensitive DNA polymerase would require it to be added after each heating step, since it would be denatured by the heat. Since Taq polymerase is stable at high temperatures, it can catalyse DNA polymerisation through the several cycles of heating involved in the PCR without fresh enzyme having to be added at each cycle – making the process simpler and cheaper. The major revolution in molecular biology which occurred during the 1990s would not have been possible without this contribution from a humble extremophile.

5   Cold Lazarus

In *Cold Lazarus*, the last play written by British television playwright Dennis Potter just before his death from cancer in 1994, Daniel Feeld awakens after 400 years into a world ruled by media moguls. Feeld, who was the central character of Potter's previous and linked play *Karaoke*, has died of cancer but his head has been frozen. He is revived by a media baron who turns Feeld's memories of twentieth-century life into a profitable nightly entertainment. The parallels with Potter's own situation are obvious and both *Karaoke* and *Cold Lazarus* are partly concerned with the relationship between playwrights and their audience. The literary device that Potter used in *Cold Lazarus* – the use of suspended animation, in this case by freezing, to propel a character into some sort of imagined future – is an old one in science fiction. However, it finds some sort of present reality in the cryogenics movement. For a mere US$28 000–$120 000, you can have your body, or for a lesser amount just your head, frozen in the hope that at some point in the future it will be possible for you to be revived and resume your existence – which only goes to show that you *can* take it with you! There are presently over 70 bodies and heads held in cryonic suspension by the various organisations in the USA that provide these facilities.

It might be worth considering what problems would have to be overcome for these bodies to be successfully revived. Not only would whatever killed them in the first place have to be remedied, the damage caused by the freezing process itself would need to be repaired. During freezing, the formation of millions of ice crystals in the body's tissues pushes their cells apart and ruptures blood vessels, which means that, when thawed, the tissues turn to mush. If the heart and lungs had not stopped already, freezing would cause them to cease, starving the tissues of oxygen and food. The formation of ice in the body fluids concentrates the salts in the water that remains unfrozen as water is sequestered into the growing ice crystals. This creates an osmotic gradient which results in the withdrawal of water from cells, resulting in their collapse and, if dehydration is sufficient, destruction (Figure 5.1). The cryogenic technicians attempt to mitigate these effects by keeping heart and lung functions going after death, by slow freezing and by perfusing the bodies with various chemicals. Despite these precautions, considerable damage must occur.

Although we may treat the claims of the cryogenecists with some scepticism, there are, remarkably, a number of organisms that have solved these problems and are able to survive the freezing of a substantial proportion of their water. These include a number of animals – some nematodes, molluscs, earthworms, insects, other arthropods and even frogs and turtles.

COLD BIOLOGY

Cryobiology is the study of the response of organisms, cells and tissues to cold, especially extreme cold and freezing. There have been two main strands to this field. The first is concerned with attempts to preserve biological materials by freezing (cryopreservation). This includes the storage of human tissues and organs for medical purposes, preserving sperm and eggs of domestic animals for breeding, banking of seeds and other material for conservation, and the preservation of food. Cryopreservation often involves exposing the material to freezing conditions that the organism, or the organism from which it originated, does not naturally survive and may not even experience under natural conditions. The other strand in cryobiology has been the study of organisms that have a natural ability to survive freezing conditions. These two strands overlap. Cryopreservation techniques are often developed largely by a method of trial and error. The techniques that prove successful are often those by which some organisms naturally survive freezing. This includes the control of the rate at which freezing and thawing occurs and the use of sugars or polyols (sugar alcohols, such as glycerol) as cryoprotectants. If we can further understand how some organisms survive freezing naturally, we can use that knowledge

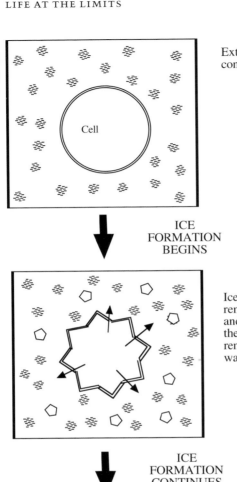

Extracellular water
containing solutes

(❄)

ICE
FORMATION
BEGINS

Ice formation
removes water
and concentrates
the solutes in the
remaining liquid
water.

ICE
FORMATION
CONTINUES

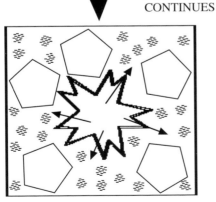

The increase in
osmotic concentration
outside the cells
draws water out of
the cells

(⟶)

FIGURE 5.1 The freeze concentration effect.

to improve our techniques for the cryopreservation of material that does not naturally survive. This chapter focusses on the ability of some organisms to survive exposure to subzero temperatures that they experience as a normal feature of their environment.

One of the first descriptions of a whole organism surviving freezing was made by Henry Power in 1663. He took a jar of vinegar that was infested with 'minute eels' (probably the vinegar eelworm *Turbatrix aceti*, a nematode) and immersed it in a freezing mixture of salt and ice. When the vinegar was thawed out 2–3 hours later, the little animals 'danced and frisked about as lively as ever'. Shortly after this (in 1683), the physicist Robert Boyle published his observations on the physical, chemical and biological effects of cold. Boyle's biological work had been inspired by the observation that bodies which had been buried in the frozen soil of Greenland were preserved for 30 years or more without any signs of putrefaction. He showed that low temperatures could help preserve eggs, meat, fruit and a variety of other biological materials, but that their texture was changed if they froze and then were thawed. He tried freezing frogs and small fish (gudgeons) in jars of water. They could revive after being encased in ice for a short period, but fish that had been frozen for three days died. The French scientist and entomologist René-Antoine Ferchault de Réaumur was the first to describe freezing experiments on insects (in 1736). He used accurate thermometers to measure the temperature of the insects, using the temperature scale which he invented (the Réaumur scale which takes the freezing point of water as zero and its boiling point as 80 °R). He found that an unnamed species of caterpillar could survive freezing to $-20\,°C\,(-17\,°R)$ and that the blood (haemolymph) of different caterpillars froze at different temperatures. He equated this to brandies of different strengths, since a weak brandy will freeze at a higher temperature than would a stronger brandy. This was the first suggestion that the resistance of animals to cold might depend on the physical and chemical properties of their blood.

Nineteenth-century travellers to the Arctic often returned with tales of frozen animals. Sir John Ross, who, together with his nephew

James Clark Ross, located the magnetic north pole, found caterpillars that had survived temperatures between −11 °C and −47 °C. Sir John Franklin, who died attempting to locate the Northwest Passage, recorded that a fish (a carp), which had been frozen for a day and a half, thawed out and leapt about enthusiastically. One explorer even described how Alaskan blackfish (*Dallia pectoralis*) kept frozen in blocks of ice for weeks were fed to his dogs. The fish apparently thawed out and revived in the warmth of the dogs' stomachs, causing them to vomit up the live fish. Unfortunately, this report has never been substantiated. These accounts, however, encouraged the belief that almost any animal could survive freezing, an idea supported by laboratory experiments which have reported the survival of snails after freezing to −130 °C, frogs to −28 °C and goldfish to −15 °C. Other reports, however, have disagreed, finding that the same species of snail could not survive −5 °C and that frogs died at −1.8 °C.

The reason for the discrepancies between these reports is that it can take a remarkable length of time for an animal to freeze. If an animal is transferred from room temperature (say 20 °C) to a freezer (at −20 °C), it will take many hours for it to cool by the 40 °C required to reach the temperature of its new surroundings. If the animal is an endotherm (such as a mammal), it can generate its own heat which, together with its insulation, delays cooling even more. Hence the reports, which have appeared in newspapers, of families' pet hamsters emerging unscathed from the freezer having been accidentally trapped there overnight. An animal could be encased in ice for some time and yet its body remain above its freezing point. Even if the animal starts to freeze, ice formation will occur in its surface layers and it may take many more hours before fatal freezing of the deeper tissues occurs. It was many years (not until 1982) before it was conclusively demonstrated that a few species of terrestrial vertebrates (reptiles and amphibians) could indeed survive freezing, with ice formation in their bodies going to completion. There have been no convincing experiments that have shown freezing survival by a mammal, bird or fish. There are, however, many invertebrates that can tolerate the freezing of their bodies.

Living in the freezer

As we saw in Chapter 2, there are a number of environments in which organisms are exposed to temperatures below 0 °C and thus the risk of freezing. In polar regions, terrestrial organisms are exposed to freezing temperatures for most of the year. In more temperate regions, they have to tolerate several months of winter, when subzero temperatures may persist for long periods of time. High mountains are another place where there is permanent snow and ice, even on the equator. Exposure to subzero temperatures may occur on a daily and/or seasonal basis.

Endothermic animals (birds and mammals) can stop their bodies from freezing by generating their own heat. They retain heat because of the insulation provided by feathers or fur, and the layer of fat beneath the skin. Other heat conservation measures include huddling together and recovering heat from exhaled breath and from the blood circulating to the extremities of the body. The Emperor penguin is perhaps the most spectacular example of these methods for survival in the cold (see Chapter 2). Endotherms can remain active in the cold if they can find enough food, or they can reduce their metabolism and lie dormant until warmer conditions return. Although air temperatures may be low, the temperature beneath an insulating layer of snow, under the ground or at the bottom of a lake may remain above 0 °C. Organisms may avoid the cold altogether by migrating somewhere warmer during the winter. Most organisms, however, cannot avoid the freezing temperatures and, for them, the choice is to survive ice formation within their bodies or to prevent their bodies from freezing.

To freeze or not to freeze?

Organisms run the risk of freezing at temperatures that are below the melting point of their body and cell fluids. There are two main responses: they can either survive ice forming within them (they are freezing tolerant) or they have mechanisms which ensure that their fluids remain liquid at temperatures that are below the freezing point of water and the melting point of their body fluids (they are freeze avoiding). The strategy that an organism uses depends on the structure and physiology it has developed during its evolutionary history and on the particular demands of its environment. If the organism is living in a wet or damp environment, ice is likely to make contact with its surface when its surroundings freeze. This may cause its body fluids to freeze by the ice travelling across the cell or body wall, or through body openings such as the mouth or anus – a process known as inoculative freezing. Most organisms surviving low temperatures in such environments are thus likely to be freezing tolerant, since inoculative freezing will cause their bodies to freeze. Some, however, may have a structure such as a cuticle, eggshell, cocoon or sheath which allows them to prevent inoculative freezing by acting as a barrier to the spread of ice into their bodies. This allows them to maintain their body or cell fluids as a liquid, despite their surfaces being in contact with external ice, and enables them to avoid inoculative freezing. In a situation where the organism is likely to be exposed to subzero temperatures with little or no water in contact with its surface (many terrestrial insects for example), it does not have the problem of inoculative freezing and it is perhaps easier for it to maintain its body fluids in a liquid state at low temperatures and thus survive by avoiding freezing.

The two strategies of cold survival are, however, not always mutually exclusive. There have been a few reports of insects which were apparently freezing tolerant switching to being freeze avoiding. The overwintering larvae of a beetle (*Dendroides canadensis*) from northern Indiana, when studied by John Duman and Kathy Horwarth from the University of Notre Dame in the winters of 1977–1978 and 1978–1979, froze at −8 °C to −12 °C but survived down to −28 °C. When examined again in 1982, however, they froze and were killed at −26 °C, apparently switching from a freezing-tolerant to a freeze-avoiding strategy during the intervening years. There are adaptations in common between freeze-avoiding and freezing-tolerant insects which may make it easy to switch between the two strategies. It must be said, however, there has been only one other report of an insect (another beetle, *Cucujus clavipes*) displaying a shift in strategy of this sort. The Antarctic nematode *Panagrolaimus davidi* is freezing tolerant when immersed in water, but, when free of surface water, there is, of course, no inoculative freezing and it can survive by avoiding freezing. The cold tolerance strategy displayed thus depends on the particular characteristics of the animal's microenvironment. The situation where the nematode is free of surface water but not desiccated is, however, likely to be a temporary one and, should water loss continue, the nematode will survive anhydrobiotically.

Many organisms cannot survive subzero temperatures at all and are killed by the lethal effects of low temperatures which are not the result of freezing. These organisms are referred to as being chilling intolerant and as suffering prefreeze mortality. The lethal effects of low temperatures include disruption of the structure of membrane and proteins, and the production of fatal imbalances between metabolic pathways. This sort of damage may only manifest itself after the organism has been exposed to low temperatures for some time. Care must therefore be taken to determine that an organism does not suffer prefreeze mortality before concluding that it can survive subzero temperatures by a freeze-avoiding strategy.

AVOIDING FREEZING

If a small quantity of pure water is cooled, it does not freeze at 0 °C and the temperature may fall to as low as −39 °C before the formation of ice crystals commences. This phenomenon is known as supercooling, which refers to the water remaining liquid at temperatures below its melting point. The temperature at which freezing eventually occurs is known as the temperature of crystallisation (or supercooling point). For freezing to occur, the water molecules need to come together to form an ice crystal that, once formed, results in the freezing of the whole body of water. This process is called nucleation and substances that cause the freezing of the water (such as an ice crystal or some other sort of particle) are called ice nucleators. The chances of the water molecules coming together spontaneously to form an ice crystal (which then acts as an ice nucleator) depends on the volume of water present, its temperature and the period of time it has been at that temperature. Substances other than ice (such as a dust particle) can act as ice nucleators and are usually responsible for initiating freezing in most natural situations.

You will rarely see supercooled bodies of water (lakes, ponds and puddles) in nature since there are usually plenty of nucleating agents (in the soil, sediment or dust) that will initiate the formation of ice and prevent supercooling. However, a small organism, which contains a small volume of water, can supercool by a substantial amount if it can prevent nucleation. This enables it to survive exposure to low temperatures by avoiding freezing.

Freeze-avoiding insects

There have been far more studies on the cold tolerance of insects than of any other group of organisms. This is due to the economic importance of insects as agricultural pests and the desire to study the survival of overwintering populations of insects, the numbers of which determine the likely size of the pest problem in the following growing season. The main pioneer of the study of insect cold tolerance was Reginald Salt, a Canadian entomologist who was inspired by the cold Canadian winters and the possibility that insect pests could be simply and cheaply controlled by exposing them to low temperatures over the winter. Salt was also impressed by being shown a technique that would allow the freezing of an insect to be measured. As he said:

> A simple laboratory demonstration . . . of the heat of crystallisation given off by a freezing insect made a deep and lasting impression on me.

Insects are enclosed by a tough and opaque cuticle. It is difficult to tell whether they are frozen or not by touching or looking at them. The technique that made such an impression on Salt solved this problem and made possible the studies which he pursued for much of his career and the studies of investigators who followed him.

The technique used by Salt employed a thermocouple to monitor the temperature of an insect. A thermocouple consists of wires of two different metals that are fused together at both ends to form a circuit, the voltage of which varies as a function of temperature. The measuring junction of a thermocouple can be made quite small (less than 1 millimetre in diameter), which allows the temperature at the surface of even a tiny insect to be measured. When water is cooled, it often does not freeze at 0 °C but cools further (supercooling) until the formation of ice crystals commences. If there is sufficient water, freezing elevates the water temperature to 0 °C, since the change in state of the water from a liquid to a solid causes it to release energy in the form of latent heat of crystallisation. The temperature of the water will remain at 0 °C until the freezing process is completed and the heat generated dissipated and the temperature can then fall to that of the surroundings. If the temperature of an insect is monitored continuously using a thermocouple, the production of latent heat during freezing can be detected. For a small insect which supercools several degrees below 0 °C, and which contains only a small amount of water, the freezing event is detected as a blip on the thermocouple trace (Figure 5.2). This allows the temperature at which the insect froze to be determined.

The supercooling point of an insect is usually measured during cooling at a constant rate of 1 °C per minute and this is often taken to be the lower lethal temperature – the temperature below which the animal will not survive. This cooling rate is, however, much faster than those occurring in nature. Slower cooling rates are likely to produce freezing at higher temperatures, since the animal spends longer times at subzero temperatures. The longer an insect is exposed to low temperatures the greater is its risk of freezing. There may also be lethal effects of low temperatures that occur before the onset of freezing. The measurement of the supercooling point of an insect cooled at 1 °C per minute may thus give an over-optimistic estimate of its cold tolerance.

Potential ice nucleators for an insect include ice in the external environment (via inoculative freezing), the molecules which make up its own structure, food in the gut and microorganisms associated with its surface or intestine. Preventing nucleation by these sources allows

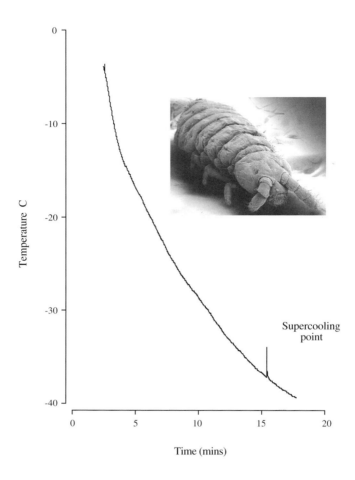

FIGURE 5.2 Freezing exotherm detected by a thermocouple attached to an Antarctic springtail (*Gomphiocephalus hodgsoni*) during cooling to −40 °C. The insect froze at −37.1 °C. Data and photo: Brent Sinclair.

the insect to increase its ability to supercool and hence to avoid freezing and survive. Inoculative freezing is avoided by overwintering in dry sites, by having a cuticle which repels water or by producing a cocoon, or some other structure, which prevents nucleation from external ice. Nucleators in the gut could perhaps be avoided by simply ceasing to feed and emptying the gut at the onset of winter, although it may be difficult to remove all nucleators by this means. By removing sources of ice nucleation, an insect could supercool to perhaps −20 °C. To survive to lower temperatures, further adaptations are required.

Freeze-avoiding insects produce a variety of small molecules, mainly sugars or sugar alcohols (polyols), which act as antifreezes. These substances include trehalose, glucose, fructose, glycerol, sorbitol, mannitol and even ethylene glycol, which is the antifreeze that you add to your car radiator. These reduce the melting point and supercooling point of the insect by their colligative (water-binding) properties, which depend on the number of ions or molecules in solution – so that their effect is directly related to their concentration. By taking up space in the solution, the relative amount of water is reduced, which has the effect of lowering the melting and supercooling points. Some insects accumulate these antifreezes in high concentrations, so that they make up as much as 25 per cent of their body weight. An arctic fly, *Rhabdophaga strobiloides*, whose larvae overwinter in willow cone galls, produces high concentrations of glycerol during winter, depressing its supercooling point to −56.1 °C. Summer larvae, which contain low concentrations of glycerol, freeze at −26.5 °C. Seasonal patterns in the production of these antifreezes have been demonstrated in many species of insects, with levels increasing during autumn and declining during spring.

As well as these low molecular weight antifreezes, freeze-avoiding insects also produce proteins which act as antifreezes. Called thermal hysteresis proteins or antifreeze proteins, they have similar properties to the proteins discovered in the blood of Antarctic fish. They prevent the growth of small ice crystals by attaching to their surface and hindering the attachment of water molecules to the ice crystal lattice. The properties of these proteins will be considered further later in this chapter when we look at the freeze-avoidance mechanisms of polar fish. The supercooled state of a freeze-avoiding insect is highly unstable and the insect may freeze and die if there is a drop in temperature below its supercooling point or if it comes into contact with an ice nucleator. Both antifreeze proteins and antifreezes such as glycerol may help to stabilise the situation by reducing the chances of ice

```
┌─────────────────────────────┐
│      FREEZE-AVOIDING        │
│         INSECTS             │
│ prevent ice forming in their bodies │
└─────────────────────────────┘
```

┌─────────────────────────┐   ┌─────────────────────────┐
│   ICE NUCLEATION   │   │   ANTIFREEZE   │
│   prevented by: │   │   MEASURES   │
│   │   │   │
│ • emptying the gut   │   │ • small molecule   │
│   │   │   antifreezes (e.g. glycerol) │
│ • avoiding surface   │   │   │
│   moisture   │   │ • antifreeze proteins inhibit │
│   │   │   ice crystal growth   │
│ • masking/removing   │   │   │
│   ice-nucleating agents │   │ • partial desiccation   │
└─────────────────────────┘   └─────────────────────────┘

FIGURE 5.3 Some of the adaptations involved in the survival of subzero temperatures by a freeze-avoiding insect. Insert drawing of a springtail by Jo Ogier.

nucleation. During entry into winter, many insects partly desiccate. The resulting loss of water will increase the concentration of antifreeze components and thus improve the cold tolerance of the insect. There are thus a suite of adaptations involved in freeze avoidance by insects, and these are summarised in Figure 5.3.

Terrestrial Antarctic arthropods consist mainly of mites (arachnids) and springtails (collembola). There are just two species of higher insects, both of them chironomids (*Belgica antarctica*, found on the west coast of the Antarctic Peninsula and its associated islands, and *Eretmoptera murphyi*, an accidental introduction to Signy Island from South Georgia). The larval chironomids are freezing tolerant, but all the mites and springtails so far examined avoid freezing by supercooling and provide good examples of this method of surviving low temperatures. Professor Bill Block and his coworkers from the British Antarctic Survey, working on the mite *Alaskozetes antarcticus* and the springtail *Cryptopygus antarcticus* at Signy Island off the coast of the Antarctic Peninsula, have investigated their adaptations. They have accumulated data from eight years of studies on the cold tolerance of these species – forming a valuable long-term database.

   Both of these animals die when they freeze but can supercool and use a freeze-avoiding strategy to survive the subzero temperatures they experience in the Antarctic (which go as low as $-26\,°C$ in extreme winters). The supercooling points show distinct seasonal patterns, with average values ranging from $-6\,°C$ in summer to $-30\,°C$ in winter. The seasonal change is due in part to the production of sugars and polyols as antifreezes. The springtail produces mainly sorbitol and mannitol, while the mite produces predominantly glycerol. The animals cease feeding at the onset of winter, thus emptying the gut of potential ice nucleators from gut-associated microorganisms and food. *A. antarcticus* produces an antifreeze protein that aids supercooling by inhibiting the growth of ice crystals. This may be important in preventing inoculative freezing since the mites can survive being encased in ice. The cuticle of *C. antarcticus* repels water, which prevents it from being trapped in water and ice. Both *Alaskozetes* and *Cryptopygus* can tolerate some desiccation with the former tolerating water loss that produces a 60 per cent reduction in its fresh weight. Desiccation may enhance cold tolerance by triggering the production of antifreezes, by reducing the amount of water and by increasing the concentration of protective compounds.

Freeze avoidance in other invertebrates
Many insects are fully terrestrial animals which, in most cases, can overwinter out of contact with water. They can largely avoid the problem of inoculative freezing that results from contact with the ice that forms in their environment. They can thus supercool and use a freeze-avoiding strategy. Invertebrates that live in the soil, or in small streams and ponds, which may partly or completely freeze during the winter, are more likely to be exposed to subzero temperatures in contact with water and to have a greater risk of inoculative freezing. Although some nematodes, for example, can survive periods of desiccation (by anhydrobiosis), they are essentially aquatic organisms and require at least a film of water surrounding soil particles for them to be active, grow and reproduce. If they are to use a freeze-avoiding strategy,

they need a structure that prevents inoculative freezing from ice crystals forming in their environment.

The potato cyst nematode (*Globodera rostochiensis*) is a plant-parasitic nematode that evolved, along with its potato host, in the high Andes of South America. It has evolved cold tolerance abilities in response to the subzero temperatures it experiences in its native mountain environment. The nematode has been accidentally spread by humans, along with the introduction of potatoes to many parts of the world. After mating, the female potato cyst nematode becomes full of eggs. She then dies and her body wall becomes hardened to form a cyst, which protects the enclosed eggs – she truly lays down her life for her children. The eggs develop into infective larvae that do not hatch until they are stimulated to do so by chemicals released by the roots of a potato plant growing nearby.

The cysts contain up to 500 eggs and can contain viable larvae after many years within the soil. The eggs are resistant to desiccation and low temperatures. This is why this nematode is so hard to control. If it infests a commercial crop, potatoes cannot be grown on that land for many years. The eggshell, which encloses the infective larva, prevents inoculative freezing, enabling the larva to supercool to temperatures as low as $-38\,°C$, even though the cyst and eggs are encased in ice. The eggs supercool to such a large extent because they contain a small volume of water with no nucleators and because the egg fluid contains trehalose that acts as an antifreeze. Other nematodes that are enclosed as an embryo or larva within an eggshell or by a sheath (derived from a partly moulted cuticle) can also prevent inoculative freezing and use a freeze-avoiding strategy.

Invertebrates that can survive anhydrobiotically (some nematodes, rotifers, tardigrades, arthropod larvae) can avoid freezing since, in a state of anhydrobiosis, there is no water present to freeze. When a soil animal prevents inoculative freezing or freezes close to, but not in contact with, ice, it will undergo dehydration. This occurs because the vapour pressure (which is a measure of the tendency of a liquid or solid to change into the gaseous state and thus evaporate) of the supercooled water within the animal is higher than that of the surrounding ice. This happens in earthworm cocoons. Earthworms are restricted to soils that contain sufficient moisture and organic material. The cocoon contains one to 20 eggs (depending on the species), which are surrounded by a tough capsule. The wall of the cocoon prevents inoculative freezing, but is surrounded by ice, and the cocoon's contents thus lose water and it effectively starts to freeze dry. This process was first described by Martin Holmstrup of the National Environmental Research Institute at Silkeborg, Denmark, who considers it to be a separate mechanism of cold hardiness and calls it the 'protective dehydration mechanism' of cold hardiness. This also occurs in some springtails (a group of insects) and enchytraeids (annelids, related to earthworms). It may also occur in other soil or aquatic invertebrates (such as nematodes) which have structures that prevent inoculative freezing.

Freeze-avoiding strategies thus represent a group of mechanisms including: supercooling in the absence of external ice (fully terrestrial species), supercooling in contact with ice outside the body (by preventing inoculative freezing), freeze drying (the 'protective dehydration mechanism') and anhydrobiosis (no freezable water present).

Polar fishes

The cold waters of the oceans of the Arctic and those surrounding Antarctica have sea ice present for much of the year. The salts dissolved in seawater result in it freezing at $-1.8\,°C$ to $-1.9\,°C$. The blood of teleost (bony) fish, however, is more dilute than seawater, reflecting their evolution in freshwater and brackish waters. The lower concentrations of salts in their blood means that teleosts freeze at $-0.6\,°C$ to $-1.0\,°C$, and we might expect those that live in polar waters to do the same. How, then, can they survive without freezing in polar waters, where the temperature is often below the freezing point of their blood? In order to do so, they need to supercool by $-1.3\,°C$ to $-0.9\,°C$, which seems a rather modest amount in comparison with the supercooling capacities of some terrestrial insects. Ice crystals are, however, ubiquitous in coastal and surface polar waters. Apart from solid masses of ice in the form of pack ice, pancake ice and icebergs, the surface waters glitter with ice crystals in suspension – frazil ice consisting of floating needles and platelets. Fish will come into frequent contact with ice crystals and, as they are supercooled, are at constant risk of freezing by being nucleated by ice crystals coming into contact with the surface of their skin, their gills and even ingested along with their food.

Fish that inhabit deep polar waters can avoid freezing by supercooling. They are able to supercool because deep waters contain no ice crystals. Ice has a lower density than seawater and hence ice crystals tend to float to the surface. Ice crystals also tend to melt in the slightly warmer deeper waters. However, the temperature is often still below the freezing point of the fish, as was demonstrated in an experiment by Scholander in the 1950s. A fjord cod (*Boreogadus saida*) can be kept alive in a tank of seawater at −1.5 °C for a long time. If a cube of sea ice is added to the water, it melts and breaks up since the temperature is above the melting point of the seawater. The ice crystals coming into contact with the fish cause it to freeze, by inoculation, and the fish dies. Deep-sea fish are only able to survive in a supercooled state since they do not come into contact with ice crystals, which would cause them to freeze.

The capelin (*Mallotus villosus*) is a small fish that is one of the most important of commercial fish, caught in large quantities for the production of fish meal and oil. It is found in Arctic waters around the world. It lays its eggs in gravel, stones and seaweed close to the shore and, in some places, even on wave-washed beaches. Here, the fish are exposed to low air temperatures and, since they only survive in moist sites, to freezing in contact with external ice. The egg can supercool to temperatures as low as −11 °C even when surrounded by ice, since its eggshell (chorion) prevents inoculative freezing. John Davenport, Director of the University of London's Marine Biological Station on the Isle of Cumbrae, showed that even a minute hole in the chorion breaches the barrier and allows the nucleation of the egg contents by ice crystals. He even observed a larval fish hatching from its egg as it was being cooled, with the tip of the fish's tail protruding through a

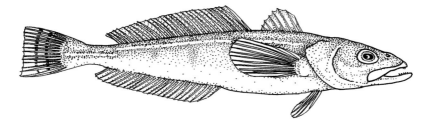

FIGURE 5.4 The Antarctic cod (*Dissosticus mawsoni*), a member of the superfamily Notothenioidea. This fish is about 1.2–1.5 metres in length. Drawing by Jo Ogier.

small hole in the chorion. The fish froze, tail first, when it came into contact with ice.

Polar fishes that live in coastal waters cannot avoid contact with ice. It is present at the surface, on the bottom in shallow waters and as floating crystals in the water column. How can these fish survive in waters which are colder than the melting point of their blood and where they are at constant risk of freezing as a result of contact with ice crystals? Although there are nearly 20000 species of teleost fish, only 274 are found south of the Antarctic Polar Front. Nearly 95 per cent of fish caught in Antarctic coastal waters belong to the superfamily Notothenioidea, the Antarctic perches (Figure 5.4). The low diversity of Antarctic fish may be partly a reflection of their isolation but may also be due to the limited number of species that are able to solve the problems of living in Antarctic waters. Professor Arthur DeVries, now at the University of Illinois–Urbana, first obtained a clue as to how the Notothenioid fish do so. DeVries has been visiting Antarctica for nearly 40 years – during the period of 1969–1998, he only missed two field seasons. The clue he found early in his career was from the results of freezing different components of the blood of Notothenioid fish.

The blood of Notothenioids melts at $-0.8\,°C$, but does not freeze until $-2.0\,°C$. There is thus a difference between the melting and freezing points (a phenomenon known as 'thermal hysteresis'). Since the temperature of Antarctic waters does not fall below $-1.9\,°C$, it remains above the freezing point of the blood and this means the fish can survive indefinitely without freezing. Since the blood of non-polar teleosts does not show this property of thermal hysteresis, there must be something special about the blood of Antarctic Notothenioids.

The molecules dissolved within the plasma of the blood of Notothenioids can be separated into small and large components, by passing them through a membrane that allows the passage of small but not large molecules. The resulting solution containing small molecular weight components, mainly salts such as sodium chloride, freezes and melts at the same temperature as the melting point of the whole blood, $-0.8\,°C$. The thermal hysteresis is due to large molecules in the blood plasma known as glycoproteins (a protein linked to a carbohydrate). These antifreeze glycoproteins comprise as much as 4 per cent by weight of the blood of the fish.

The antifreeze glycoproteins of Antarctic Notothenioids act in a different way to the low molecular weight antifreezes, such as glycerol, which are found in freeze-avoiding insects. The antifreeze proteins of fish attach to the surface of ice crystals and inhibit their growth by preventing further water molecules from joining the ice crystal lattice. If an ice crystal is introduced into a purified solution of antifreeze protein, at a temperature between its melting and freezing points, it will remain but not grow even after a week. Fish antifreeze proteins do not just slow the growth of ice crystals, they stop it completely. The freezing of the supercooled fish is thus prevented since any ice crystals that find their way into their bodies are immediately coated with antifreeze protein. If the temperature is lowered sufficiently, the inhibition will be overcome and an ice crystal will grow. Antifreeze proteins lower the freezing point of the fish by just a couple of degrees but this is enough to enable them to survive in an environment where the temperature does not fall below $-1.9\,°C$.

Antifreeze proteins are present not just in the blood of the fish but also in the fluids of the body cavity and those of the heart, liver and muscles. These proteins are produced in the liver and circulated around the body. The brain is the only organ in which antifreeze proteins have not been found but this is surrounded by an extradural fluid which does contain them and which presumably acts as a barrier to ice formation. The gut is a particularly important site of potential ice nucleation since ice crystals are ingested along with the food. Antifreeze proteins are secreted into the gut via bile and there is evidence that they may have evolved from gut enzymes (see Chapter 8). There are no antifreeze proteins in the urine since they are filtered out by the kidneys, a process which retains the proteins within the blood.

Just about all Antarctic teleosts contain antifreeze glycoproteins. Although the Antarctic fish fauna is dominated by Notothenioids, there are fish from other families such as eel pouts (Zoarcidae) and icefish (Channichthyidae) that contain them. Arctic fish from a number of families also contain antifreeze proteins. These include Arctic and Greenland cod (Gadidae) and winter flounder (Pleuronectidae). These proteins have evolved independently in a number of different groups of fish to solve the problems of survival in icy waters. This is a striking example of convergent evolution, which I will tell you more about in Chapter 8. The antifreeze proteins of many Arctic fish appear on a seasonal basis, only developing significant concentrations during the winter. The production of antifreeze proteins in these fish is triggered by both decreasing temperatures and shorter day lengths and is under hormonal control.

Since their discovery in Antarctic fish, antifreeze proteins have also been found in freeze-avoiding insects. Here, they play a similar role by preventing the growth of ice crystals and thus stabilising the super-cooled state. Antifreeze proteins are also called thermal hysteresis proteins in insects. They often have a much greater hysteresis activity than the antifreeze proteins from fish, producing a difference between the melting and freezing point which may be as much as 11 °C. These proteins have been reported from a variety of insects (although beetles feature most prominently) and from spiders, centipedes and mites. Antifreeze proteins have also been reported from freezing-tolerant insects and from other freezing-tolerant invertebrates (nematodes and molluscs). Their role may be rather different in this situation, as I will describe later, and these may turn out to be a rather different group of proteins. Antifreeze proteins have also been discovered in plants.

Polar fish have had to solve other problems which result from exposure to low temperatures, apart from the risk of freezing. Their enzymes, membranes and physiological processes have to function at temperatures which are constantly at about $-1.9\,°C$. Liquids tend to become more viscous or sticky at low temperatures; try keeping your bottle of golden syrup in the fridge, it makes it harder to pour. The fish's heart needs to pump blood around the body to distribute food and oxygen to the tissues. The low temperature tends to make the blood more viscous and harder to move. This is offset by the fact that more oxygen will dissolve in the blood at low temperatures than will at high temperatures. The fish can thus make their blood less viscous by reducing the numbers of red blood cells, and hence the amount of haemoglobin (which carries oxygen), and yet still supply the tissues with enough oxygen. Icefish (Channichthyidae) have carried this to its ultimate conclusion. They have no haemoglobin or red blood cells, earning them the name white-blooded fish, and all their oxygen is carried dissolved in the plasma or fluid of the blood. Their blood can carry only 10 per cent of the oxygen transported by an equivalent volume of blood from a red-blooded fish. They compensate for this by having larger hearts to circulate a greater volume of blood and by having more blood vessels in their gills and skin to extract oxygen from the seawater.

TOLERATING FREEZING

In some respects, organisms that survive subzero temperatures by avoiding freezing and supercooling are living dangerously. The supercooled state is unstable. Fatal freezing may result from contact with an ice nucleator, exposure to temperatures below their supercooling point or to temperatures below those at which their antifreeze proteins can prevent ice crystal growth. Since the process of freezing may be the most stressful event associated with subzero temperatures, the ability to tolerate freezing (rather than avoiding it) should provide a less risky solution to the problems of living in very cold environments. A number of animals can tolerate ice forming in their bodies, to varying degrees, including some insects and other arthropods, molluscs, nematodes, earthworms, frogs and turtles.

How much ice must form in the body of an animal before we can consider it to be freezing tolerant? Most animals will survive some degree of ice formation in their bodies. Even we humans will survive some parts of our bodies freezing, although we will lose the tips of our fingers or toes to frostbite if they freeze. Freezing kills our tissues and we die if more than a small proportion of our bodies freeze. Freezing-tolerant animals will survive the bulk of their body water turning to ice. In fact, some can survive the freezing of all their freezable water. Not all cellular water can be frozen, at least at temperatures likely to be encountered in the environment. A proportion of the water is restricted in its movement by interactions with membranes, proteins and other macromolecules or structures in the cell. This is called 'osmotically inactive water' since it is not free to move under an osmotic stress. Osmotically inactive water is sometimes also called 'unfreezable water', since it does not freeze (at least at temperatures down to $-45\,°C$), or 'bound water', since it is attached to cell molecules and structures. The term 'bound water', however, is misleading since it can rapidly exchange with the free water in the cell. Osmotically inactive water constitutes about 18 per cent of cell water and hence the maximum amount of ice formed in freezing-tolerant animals is about 82 per cent of total body water. Freezing-tolerant animals survive the freezing of 50–82 per cent of their water.

Measurements of freezing survival in the laboratory, however, can be deceptive and we must always bear in mind the conditions to which organisms are exposed in their natural environment. This is particularly important for groups, such as reptiles, in which the ability to survive freezing is poorly developed. There has been considerable debate as to whether hatchlings of the painted turtle (*Chrysemys picta*), and other turtle species, are freezing tolerant or avoid freezing by supercooling. Freezing tolerance is considered to be relevant to the biology of the animal when it will survive continuous freezing for days or weeks at temperatures usually encountered in its environment and with its ice content reaching a constant level.

It is generally thought that animals only survive ice forming in their bodies if it is confined to their body cavities and other extracellular spaces. Freezing of their cells (intracellular freezing) is thought to be fatal. Intracellular freezing is partly prevented by the animals freezing very slowly, over a period of days in some cases. At least one nematode, however, breaks all the rules. As we will see later, it freezes in a matter of seconds and survives extensive intracellular freezing.

Freezing-tolerant insects

There are more than 60 species of insects which are freezing tolerant, surviving the freezing of their tissues and body fluids. Most freezing-tolerant insects freeze at −5 °C to −10 °C but will tolerate freezing to much lower temperatures. Woolly bear caterpillars, larvae of a moth from the Canadian Arctic (*Gynaephora groenlandica*), will survive freezing to −70 °C and fruit fly larvae (*Chymomyza costata*) survive temperatures below −100 °C. Larvae of the goldenrod gall fly (*Eurosta solidaginis*) overwinter in galls on the stems of their host plant. Since the stems project above any snow and the galls provide little insulation, the larvae are exposed to very low temperatures in parts of their range. In winter, populations from the northern United States are freezing tolerant and will survive below −40 °C, with 64 per cent of their water frozen.

While many freezing-tolerant insects can survive temperatures far below that at which they freeze, some show only a moderate amount of cold tolerance. The New Zealand alpine cockroach, *Celatoblatta quinquemaculata*, freezes at about −5 °C with 74 per cent of its body water being converted into ice, but will only survive down to about −10 °C. From the same environment, living under slabs of schist rock in the mountain ranges of New Zealand, is the alpine weta *Hemideina maori*. This is an orthopteran insect (related to grasshoppers) and is the largest freezing-tolerant insect known. Like the cockroach, its cold tolerance abilities are modest. It will survive the freezing of 82 per cent of its body water at −5 °C but its lower lethal temperature is about

−10 °C. Since these insects survive ice formation in their bodies going to completion, it is not the freezing process itself which kills them but some other unknown factor at lower temperatures.

Most freezing-tolerant insects freeze at fairly high subzero temperatures. This may be important since the greater the degree of supercooling the faster and more violent will be the freezing process when it occurs. By preventing supercooling, freezing is slow and gentle, giving the cells time to adjust to the changes occurring within the body of the insect. The freezing process can be very slow indeed, with *H. maori* taking 10 hours to reach its maximum ice content at −5 °C and larvae of the goldenrod gall fly, *E. solidaginis*, taking 1–2 days at −23 °C.

Many freezing-tolerant insects produce proteins or lipoproteins in their haemolymph that act as ice-nucleating agents. Ice-nucleating agents trigger ice formation by binding water at their surface in such a way that it tends to form an ice crystal. These ice-nucleating proteins not only ensure that freezing occurs at a relatively high subzero temperature, and hence the propagation of ice is slow, but they may also prevent fatal intracellular freezing by ensuring that ice formation occurs in the extracellular fluids of the animal. As freezing proceeds in the haemolymph, the ice excludes salts and other solutes from its structure, raising their concentration in the remaining unfrozen portion (the freeze concentration effect). This creates an osmotic gradient that dehydrates the cells, raising their internal concentrations and preventing them from freezing (see Figure 5.1). Not all freezing-tolerant insects possess ice-nucleating proteins in their haemolymph, however, and some may rely on food or microbes in their gut which act as nucleating agents or on inoculative freezing by contact with external ice to produce freezing at high subzero temperatures. A few Arctic insects can supercool to below −50 °C and yet still survive freezing, when it occurs, in contrast to the normal pattern.

Many freezing-tolerant insects produce low molecular weight compounds in response to the onset of winter. These include sugars (trehalose, glucose), sugar alcohols (glycerol, sorbitol, erythritol) and amino acids (proline, alanine). Some of these will sound familiar since they are also produced by freeze-avoiding insects where they act as antifreezes. In freezing-tolerant insects, they play a rather different role, acting as cryoprotectants. Their role as cryoprotectants depends partly on their colligative (water-binding) properties. They depress the melting point and reduce the amount, and rate, of ice formation at any temperature where freezing occurs. Since cell dehydration is a major cause of cell injury, decreased ice formation reduces the cellular dehydration which occurs as a result of the freeze concentration effect. Cryoprotectants that can penetrate cell membranes (such as glycerol) will enter the cells and thus counteract the effects of cell dehydration. Some cryoprotectants protect cells by mechanisms that do not rely on their colligative properties. Trehalose attaches to proteins and membranes protecting them against the harmful effects of dehydration (as we saw in Chapter 3 on anhydrobiosis). Since freeze concentration effects result in cellular dehydration, the protective properties of trehalose may be important in tolerating freezing. Desiccation may trigger the synthesis of cryoprotectants and increase their concentration through the loss of water.

Some freezing-tolerant insects have been shown to produce antifreeze proteins. This may seem surprising since, in freeze-avoiding insects, these proteins act to inhibit ice nucleation and promote supercooling. Most freezing-tolerant insects freeze at high subzero temperatures and may promote such freezing by producing ice nucleating proteins. Antifreeze proteins play a rather different role in freezing-tolerant insects, which depends on their ability to interact with the surface of ice crystals. If you keep a body of ice frozen at a constant subzero temperature, it is not stable, particularly at high subzero temperatures. It undergoes recrystallisation – slightly larger ice crystals grow at the expense of slightly smaller ones, resulting in fewer but larger crystals. This occurs if you keep ice cream in your freezer for too long – it goes 'icy'. If this change in the size of ice crystals occurred within a frozen insect it could be quite damaging as the crystals grind against the membranes of cells. The antifreeze proteins produced by freezing-tolerant insects inhibit ice recrystallisation since they sit in the interfaces between ice crystals and prevent the exchange of water molecules between them. They can produce this recrystallisation inhibition effect even if they are present in quite low concentrations.

Antifreeze proteins from freezing-tolerant insects also show a thermal hysteresis effect (a difference between the melting and freezing point in the presence of an ice crystal), although this has no apparent function in these insects and is a consequence of the attachment of the protein to the surface of ice crystals. Both the antifreeze proteins and the cryoprotectants produced by freezing-tolerant insects might be expected to inhibit ice nucleation, to the disadvantage of the insect. These effects are, however, easily overcome by the presence of potent ice nucleators in the insect's haemolymph and/or gut, or by inoculative freezing. In some other animals (molluscs and nematodes), recrystallisation inhibition has been reported where there is little or no thermal hysteresis. The structure of the proteins responsible has yet to be determined but they may well turn out to be a new type of ice-active protein.

It is often assumed that insects will only survive freezing if the ice is confined to their haemocoel and extracellular spaces. This has rarely been tested and there are at least two insects that will survive some intracellular ice formation. Larvae of the goldenrod gall fly (E. solidaginis) accumulate cryoprotectants (glycerol and sorbitol) and promote freezing by the presence of inorganic crystals (calcium phosphate) that act as ice nucleators. Reginald Salt was the first to report that the fat body cells of E. solidaginis could survive intracellular freezing (the fat body is the insect's equivalent of the liver). These observations have been confirmed and expanded by the work of Rick Lee's laboratory at Miami University in Ohio. Bill Block, and his coworkers at the British Antarctic Survey have claimed that the fat body cells of a freezing-tolerant Arctic fly from the Norwegian archipelago of Svalbard (Heliomyza borealis) will also survive intracellular freezing. The ability to tolerate intracellular freezing may be much more common that we think. However, in my own laboratory, we were unable to demonstrate intracellular freezing in the fat body cells or Malphigian tubules of the New Zealand alpine weta (H. maori), despite the freezing of 82 per cent of its body water.

FREEZING-TOLERANT INSECTS
survive ice forming in their bodies

ICE NUCLEATION
encouraged by:

• ice-nucleating proteins in haemolymph

• food/bacteria in gut

• inoculative freezing

CRYOPROTECTIVE MEASURES

• freeze at a high subzero temperature

• prevention of intracellular freezing

• small molecule cryoprotectants (e.g. glycerol)

• ice-active proteins inhibit recrystallisation

FIGURE 5.5 Some of the adaptations involved in the survival of subzero temperatures by a freezing-tolerant insect. Insert drawing of a beetle by Jo Ogier.

The adaptations involved in the survival of freezing by insects are summarised in Figure 5.5.

Other freezing-tolerant invertebrates

Many invertebrates live in the intertidal zone, the area of the seashore that is exposed at low tide and covered at high tide. Those that cannot migrate to avoid cold when they are uncovered by the tide may have to face freezing temperatures. There are a number of intertidal invertebrates that can survive prolonged exposure to temperatures of $-10\,°C$ or lower. These animals cannot usually avoid freezing since they are in close contact with seawater or with silt and food material which is likely to cause ice nucleation, unless they live at the bottom of a deep rock pool which does not completely freeze. Most of the intertidal invertebrates that can survive low temperatures are freezing tolerant. The periwinkle (*Littorina littorea*) will survive to $-22\,°C$ with 76 per cent of its body water converted to ice and the boreal barnacle (*Semibalanus balanoides*) to $-20\,°C$ with 80 per cent of its body water frozen. Intertidal invertebrates are thought to survive only if ice is limited to the extracellular spaces of the animal. This is encouraged by the production of ice-nucleating proteins. Sugars and sugar alcohols, which are used as cryoprotectants in freezing-tolerant insects, do not appear to be used in this way by intertidal molluscs. They do, however, produce amino acids (e.g. proline, alanine, taurine and strombine) which appear to act as cryoprotectants. Antifreeze proteins are produced by the blue mussel (*Mytilus edulis*). These have little thermal hysteresis activity and function as recrystallisation inhibitors. Intertidal molluscs develop the ability to survive freezing on a seasonal basis, with the production of protective compounds triggered by falling temperatures. Freezing tolerance may also be initiated or enhanced by desiccation, increased salinities and low oxygen concentrations.

The Antarctic is a good place to look if you are interested in organisms that survive freezing. In 1989, I visited Antarctica for the first time and isolated a soil nematode, *Panagrolaimus davidi*, which I have kept in culture in my laboratory ever since. *P. davidi* was isolated at Cape Bird, Ross Island from an ice-free area where algae were growing on loose soil and shingle which received meltwater from an adjacent snowbank. Here, temperatures do not rise above 0 °C for much of the year and the nematode is likely to remain either frozen or dry for at least eight months. During the spring and summer, there is sufficient heat from the sun for liquid water from melting snow to be available. Even then, the substrate periodically freezes. When I revisited the site in the spring of 1997, it was clear that the nematode's habitat was freezing and thawing on a daily basis during that part of the year. The nematode is thus regularly exposed to freezing and has to survive inoculative freezing from the ice surrounding it.

The optimum growth temperature of *P. davidi* is, perhaps surprisingly, 25 °C and it does not grow at all at temperatures below 6.8 °C. At 25 °C, the nematode will complete its life cycle in just seven days. However, temperatures are above its threshold for growth for only short periods of the day and year. Ian Brown, who did his doctoral research on this nematode in my laboratory, has calculated that *P. davidi* is lucky to complete one generation a year in its natural habitat.

It shows resistance adaptation to the extreme conditions – surviving and lying dormant for much of the time and growing rapidly during the brief periods when conditions are favourable.

Freezing is a major stress for the nematode, although its substrate must also desiccate when sources of water, such as melting snow, become exhausted. The nematode is freezing tolerant and will survive freezing in water to temperatures as low as $-80\,°C$. Nematodes are transparent and we can observe ice forming in their bodies when they are cooled on a microscope cold stage. Freezing is obvious since the ice scatters light and the nematode suddenly darkens when it freezes. The darkening of the nematode is so complete I found it hard to imagine that the ice was confined to its extracellular spaces. In 1994, Donald Ferns, a student research assistant, and myself set out to determine where, and how, ice formed in this nematode. To do this, we used a cold microscope stage mounted on a research microscope, which allowed us to observe the freezing process at a high magnification. This was recorded on video, which could then be analysed frame by frame.

Ice formed in the medium surrounding the nematode first. Freezing of the nematode itself occurs by inoculative freezing, the ice travelling into the body through natural openings. This sometimes occurs via the mouth and sometimes the anus, but most often via the excretory pore. From here, the ice travels throughout the body and different parts can be seen to freeze at slightly different times. All compartments of the body freeze, including the intracellular compartments (Figure 5.6). Our strain of *P. davidi* is parthenogenic; females lay unfertilised eggs that develop into females. We can take a single nematode, in which we have observed intracellular freezing, and place it in culture, where it will recover, grow and lay eggs that develop into the next generation of larvae. This was the first (and so far only) example of an intact animal surviving extensive intracellular freezing, something that was previously thought to be impossible (but note that intracellular freezing has been demonstrated in insect fat body cells). We have confirmed the presence of intracellular ice using electron microscope techniques that preserve them or the spaces where they occurred (freeze fracture and

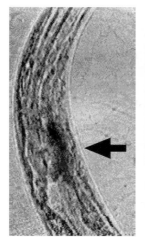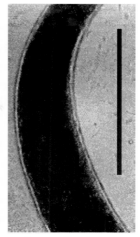

FIGURE 5.6   The freezing of the Antarctic nematode *Panagrolaimus davidi* starts near the posterior bulb of the oesophagous (seen as a darkening as indicated by the arrow). Freezing spreads throughout the body until all body compartments freeze. These pictures are taken from a video sequence of the freezing of the nematode. Scale bar = 100 μm. From Wharton & Ferns (1995), reproduced with the permission of the Company of Biologists Ltd.

freeze substitution). The ice crystals are found throughout the cells of the nematode but are restricted to the cytoplasm; ice formation within cell organelles appears to be fatal. Our studies on this are continuing.

How is this nematode able to survive in this remarkable way? The nematode survives 82 per cent of its water freezing. The freezing process is extremely rapid, with all parts of the body freezing within a few seconds. This is in marked contrast to other freezing-tolerant animals that take hours or days to complete the freezing process. Perhaps the survival of intracellular freezing requires the almost simultaneous freezing of all body compartments to avoid the osmotic stresses which would result if they froze at different times. The nematode produces trehalose at concentrations equivalent to those found in some freezing-tolerant insects, which might thus be acting as a cryoprotectant. Working with Craig Marshall of the Biochemistry Department of the University of Otago, we have recently demonstrated the presence of a protein that functions as a recrystallisation inhibitor. Recrystallisation inhibition may be particularly important for the nematode to survive intracellular ice, since recrystallisation of the ice during long-term freezing might be expected to cause extensive damage. We are planning to isolate and purify the protein involved and to determine its structure so that we can compare it with ice-active proteins from other animals.

Nematodes are essentially aquatic organisms and they are likely to undergo inoculative freezing from the surrounding water unless they have a structure, such as an eggshell or a sheath, which prevents it. *P. davidi* is the only nematode in which the survival of intracellular freezing has been demonstrated but such an ability may prove to be widespread among nematodes.

Other groups of invertebrates, which inhabit soil or freshwater sites which periodically freeze, may be subject to inoculative freezing and need to be freezing tolerant to survive. Some tardigrades from Greenland and earthworms from Siberia have been reported to be freezing tolerant.

Frigid frogs and cold turtles
Many ectothermic vertebrates (fish, amphibians and reptiles) overwintering in cold temperate environments can avoid the risk of freezing. They hibernate deep in the soil or at the bottom of ponds and lakes. Although the surface of a lake may develop a thick layer of ice, the water beneath remains liquid and, as long as the lake does not freeze completely, the temperature stays above 0 °C. There are, however, a few species of amphibians and reptiles that hibernate in shallow terrestrial sites where they freeze over the winter. They need some special adaptations in order to survive in these sites. Why have they chosen such an apparently difficult and risky option for spending the winter, compared with vertebrates that hibernate in sites that are unlikely to freeze? When spring comes, the shallow sites of these animals are likely to thaw long before those of animals hibernating in more protected sites. The animals can become active quicker and can thus start

feeding and growing. They are the first to exploit the resources that become available in the spring and are likely to be able to breed earlier and hence more successfully than their competitors.

The first demonstration that a vertebrate could survive freezing in its natural habitat came relatively recently. In 1982, William Schmid of the University of Minnesota at Minneapolis showed that three species of frog could survive freezing. There are five species of frog that have now been shown to be freezing tolerant: *Rana sylvatica* (the wood frog), *Pseudacris triseriata* (the striped chorus frog), *Hyla versicolor* (the grey tree frog), *H. crucifer* (the spring peeper) and *H. chrysoscelis* (Cope's grey tree frog). The Siberian newt (*Salamandrella keyserlingii*) also tolerates freezing, with a range that extends into the Arctic tundra. There are two species of turtle, the hatchlings of the painted turtle (*Chrysemys picta*) and the box turtle (*Terrepene carolina*), which are able to tolerate freezing. Some other reptiles, including garter snakes, some lizards and some other turtles, can survive a small amount of ice in their bodies, but they do not survive if all the water that could freeze at subzero temperatures actually does so. Their ability to survive a small amount of ice formation, however, can be critical for their survival. The eastern garter snake (*Thamnophis sirtalis*), for example, commonly encounters subzero temperatures while it is active during spring and autumn outside its winter hibernating site. It will tolerate 35 per cent of its water freezing and exposure to subzero temperatures for more than 48 hours, enabling it to survive frost episodes, but it is killed if it freezes completely. During winter, the snakes hibernate in caves underground where they are protected from freezing. Suitable sites are in short supply and they may have to migrate long distances to reach them. Garter snakes congregate in enormous numbers and you are likely to find yourself knee deep in snakes if you venture into one of these caves during winter.

The main laboratories studying freezing tolerance in frogs have been those of Ken and Janet Storey at Carleton University in Canada and of Rick Lee and Jon Costanzo at Miami University in Ohio. The wood frog (*R. sylvatica*) has been the main subject of their investigations. This frog has a wide distribution in North America, from the southeastern United States to Alaska. It generally inhabits damp woodlands and overwinters in shallow sites beneath leaf litter. These sites are damp and ice crystals can easily penetrate the skin of the frog. It will survive temperatures down to −6 °C, with 65–70 per cent of its body water frozen for four weeks or more.

The freezing of *R. sylvatica* is initiated by inoculative freezing across the skin from ice forming in its moist hibernating site. Ice-nucleating bacteria have been isolated from its gut. These may trigger ice nucleation if the surface of the frog is dry and there is no inoculative freezing. There are ice-nucleating proteins in the blood but these trigger ice nucleation at temperatures below that initiated by inoculative freezing or by bacteria in the gut. The blood of many vertebrates, including those that do not tolerate freezing, contains proteins that have some ice-nucleating properties. This is clearly not their function in these animals and it is doubtful that the ice-nucleating proteins in the blood of *R. sylvatica* play an important role in the freezing tolerance of the animal. Since their hibernating sites are damp, there are likely to be abundant ice crystals in contact with the skin to produce inoculative freezing as soon as the temperature falls below the melting point of the frog's blood, at about −0.4 °C. As in freezing-tolerant insects, freezing at a high subzero temperature means that the freezing process is slow and gentle and may ensure that there is no intracellular freezing. The freezing process takes 24 hours to complete at −2.5 °C.

Researchers have observed the freezing process in *R. sylvatica* by using proton magnetic resonance imaging (a technique used in medicine to provide images of the inside of a patient's body). Ice starts to form at one point on the surface of the skin, spreads over the surface and works its way inwards. The body cavity and the spaces between the frog's organs freeze next. This withdraws water from the organs and sequesters it into ice in the body cavity, which may help to prevent the tissues freezing intracellularly by partially dehydrating them by up to 50 per cent or more. The ice then migrates into the extracellular spaces of the tissues and organs themselves. The liver is the last organ to freeze. Once the frog is fully frozen, its trunk and limbs are rigid, its skin is covered in frost and its eyes are opaque because of the freezing of the eye fluids. Ice is found beneath the skin, between the muscles, a mass of ice fills the abdominal cavity and the organs appear shrunken.

During the freezing process, physiological and biochemical changes occur within the frog that enable it to survive. Levels of glucose in the blood increase enormously, by its release from glycogen stores in the liver once freezing has commenced. This occurs very quickly, with the concentration of glucose increasing within a matter of minutes of freezing commencing and eventually reaching 100–200 times its normal levels. The production of glucose appears to be controlled by receptors on the liver cells that are sensitive to adrenaline. Adrenaline is a hormone that is widely distributed in vertebrates and is released in response to stress, eliciting a 'fight or flight' response. One of its effects is to trigger the mobilisation of glucose from glycogen, providing the body with fuel for action. Another of its effects is to produce an increase in heart rate. This occurs in the freezing frog, with the heart rate declining as the temperature falls but increasing markedly as soon as freezing commences. The increase in heart rate could also be explained by the elevation in temperature during the freezing process (in the form of heat of crystallisation). The increased heart rate ensures that the glucose produced by the liver is distributed to the tissues, that the tissues continue to be supplied with oxygen and food during the freezing process, and may assist in the partial dehydration of the tissues. As freezing is completed, the heart rate slows and eventually the heart stops beating altogether.

Glucose is thought to act as a cryoprotectant by reducing the dehydration of cells and the amount of ice that is formed. It may also help preserve membranes and proteins in a similar manner to trehalose. The production of glucose in response to freezing occurs in the other species of freezing-tolerant frogs, with the exception of *H. versicolor*, which produces glycerol instead. High levels of glucose in the blood (hyperglycaemia) are damaging to most animals; in humans, for example, diabetes results when the regulation of carbohydrates breaks down. Perhaps the frogs can tolerate the high glucose levels since it only occurs after freezing commences and the frogs' metabolism is depressed as the freezing process proceeds.

Recently, Ken and Janet Storey at Carleton University have been using a gene-based approach to search for freeze tolerance adaptations in frogs. This involves comparing the proteins produced by frogs after a freezing stress with those produced by control non-frozen frogs. They have identified several proteins, and thus the expression of several genes, which are induced by freezing. Among these are fibrinogens, produced by the liver. Fibrinogen is a blood protein that is involved in clotting, wound healing and repair. Its production in response to freezing in frogs suggests that survival involves the repair of damage, which occurs during ice propagation through the body and/or during thawing. One advantage of this sort of screening approach is that it can reveal adaptations which were previously unsuspected. Its application to the study of cold tolerance in animals (or to desiccation survival) is in its infancy, although there has been more extensive work on plants and microbes.

During thawing, the processes that occurred during freezing are reversed. Thawing starts in the core of the body, due to higher concentrations of glucose in the organs, and spreads outwards. The resumption of heart activity is the first sign of life; it starts to beat even before the melting process is completed. This may be important to ensure that the thawing tissues are supplied with oxygen and to assist in the redistribution of glucose in the body. Glucose is rapidly converted back into glycogen in the liver, thus avoiding the stress of exposure to high glucose levels in the recovering tissues. Breathing recommences, the frog shows some sort of activity about seven hours after the start of thawing and normal behaviour resumes after 24 hours of recovery. The changes occurring in the frog during freezing and thawing are shown in Figure 5.7.

Freezing-tolerant frogs are of great interest to medical scientists who would dearly love to be able to freeze and thaw human hearts, and other organs, to store them for transplantation. The frogs seem to have solved

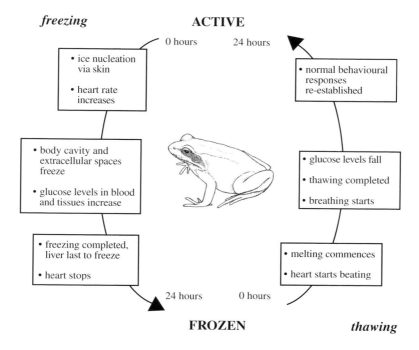

freezing   **ACTIVE**

0 hours   24 hours

- ice nucleation via skin
- heart rate increases

- normal behavioural responses re-established

- body cavity and extracellular spaces freeze
- glucose levels in blood and tissues increase

- glucose levels fall
- thawing completed
- breathing starts

- freezing completed, liver last to freeze
- heart stops

- melting commences
- heart starts beating

24 hours   0 hours

FROZEN   *thawing*

FIGURE 5.7 Some of the changes occurring during the freezing and thawing of the wood frog, *Rana sylvatica*. Insert drawing by Jo Ogier.

this problem of freezing and thawing. They can also tolerate high blood glucose levels, low levels of oxygen and the other potentially harmful changes which occur in their bodies during freezing and thawing.

Although the ability of certain frogs to survive freezing is now well established, there has been rather more debate over the role of freezing tolerance in reptiles. Painted turtles (*Chrysemys picta*) have a wide distribution in cool temperate regions of North America, being found from the east to the west coast in the northern United States and southern Canada. Eggs are laid in shallow nest cavities in exposed banks. The turtles hatch in late summer but the hatchlings do not leave the nest until the following spring. They face low temperatures and the risk of freezing in their relatively exposed nest sites over winter.

Temperatures as low as $-12\,°C$ have been recorded in their nests in the sandhills of Nebraska, where winters are particularly severe.

Reptiles are much less susceptible to inoculative freezing than are frogs since they have a skin that is tough, dry and waterproof. Turtles have a shell, which may also provide some resistance to ice inoculation. The hatchlings of the painted turtle are small, which makes it easier for them to supercool. Much of their ability to tolerate the cold winter has often been ascribed to their ability to supercool. However, they can also survive freezing and can tolerate about 50 per cent of their body water as ice. They thus have the capacity either to tolerate or to avoid freezing. They cannot do both at once, however, since their ability to tolerate freezing requires freezing at a relatively high subzero temperature (little supercooling), while they cannot survive freezing following extensive supercooling. The strategy they use depends on the microenvironment within their nest over the winter.

After freezing, the hatchlings can only survive down to about $-4\,°C$. To survive much lower temperatures (down to $-10\,°C$ to $-12\,°C$) the hatchlings must supercool. Their ability to supercool, however, is determined by the soil and moisture characteristics of their nesting site. Although turtles are less susceptible to ice inoculation than frogs, inoculative freezing from ice forming around them in their nest can still occur via the mucous membranes of the cloaca (the common opening of the digestive, urinary and reproductive systems), the nostrils or the eyes, and even via wounds in the skin. Exposure to freezing in sand, even containing as little as 2.3 per cent water by weight, results in inoculative freezing. In soil with a high clay or organic content, however, the hatchlings can supercool. Whether the turtles survive the winter by tolerating freezing or by supercooling thus depends upon an interplay between the thermal environment, the moisture content and the characteristics of the soil in its nest site.

COLD TOLERANCE IN PLANTS

If you are a gardener in an area with cold winters, you will know that most plants do not grow over the winter and survive the conditions in a dormant state. The first cold snap will kill off flowering annuals, which then overwinter as seeds. Early growth in the spring will be destroyed by a late frost and so careful gardeners will cover the emerging shoots of their early potatoes with earth to protect them.

Plants are subjected to chilling stress at temperatures in the range 0–15 °C and to freezing (frost) stress at temperatures below 0 °C. The responses of plants to low temperatures are perhaps harder to categorise than those of animals. Parts of the plant are protected below ground and plants can lose large parts of their structure but survive and recover. Plants suffer not just because of the low temperatures themselves but also because of associated problems with the availability of water, nutrients and oxygen. There is great variety in the ability of plants to tolerate low temperatures, reflecting the temperature environment of their natural habitat. Tropical and subtropical plants, including crops such as maize, soybean, tomatoes and cucumbers, suffer cold stress at temperatures below 15 °C. Even though I have a greenhouse, my tomatoes have not done well this year because of a relatively cool summer. Many plants, however, grow successfully in cold northern temperate conditions and in alpine and arctic tundra. How do they survive the winter when air temperatures can sometimes fall as low as −60 °C?

The roots and parts of the plant below ground, or covered by snow or leaf litter, are insulated to a large extent from freezing. The plant may have its own insulation in the form of a downy or woolly covering, a dense mass of dead attached leaves or a thick bark. The large mass of some plants allows them to store heat absorbed during the day and prevent freezing at night. The plant may have ways to increase heat absorption from the environment, such as a rosette form in the leaves and flowers. Some plants protect their sensitive parts by accumulating water, which takes a long time to freeze, so that the temperature is held close to 0 °C for much of a short freezing event. The plant may lose, or not grow, its more sensitive tissues (young shoots, leaves and flowers) during the winter or it may die back almost completely and overwinter as a dormant seed, bulb, tuber, corm or rhizome. Some plants even show a degree of endothermy, generating heat from metabolic activity. For example, the skunk cabbage (*Symplocarpus foetidus*) of North America can maintain the temperature of its flowers at 10 °C, even though the air temperature may be −15 °C, by increasing its rate of respiration. This plant is the earliest flowerer in the midwestern and northeastern USA, often pushing its large and foul-smelling flowers above the snow. Its elevated temperatures help it to attract insect pollinators.

Despite these various strategies, many plants spend the winter with parts of their structure above the ground and exposed to temperatures that could fall to many degrees below 0 °C. The survival mechanisms of plants are similar to those of animals. They are either freezing tolerant or they avoid freezing by supercooling. However, because they usually contain a large quantity of water and are in direct contact with frozen soil or frost, which causes inoculative freezing, extensive supercooling is rare in plants and most survive by tolerating ice forming within them.

Many plants are able to supercool and survive a brief exposure to temperatures just a few degrees below the melting point of their tissues. More extensive supercooling is, however, rare. The leaves of some evergreen plants have relatively little space between their cells, containing small volumes of water that are isolated from one another by leaf veins. This allows them to supercool to below −20 °C. In winter, the most sensitive tissue of the twigs of apple trees, the parenchymal cells of the xylem (water-conducting tissue), can supercool to −40 °C but die when they freeze. The stem tissues of temperate deciduous woody angiosperms (like the apple twigs) contain little space between the cells and so, if freezing takes place, it occurs within the cells – thus killing them. They therefore have to supercool to survive.

Plants tolerate ice forming in parts of their structures as long as their cells do not freeze. The ice may form in the extracellular spaces between the cells. This raises the osmotic concentration of the unfrozen portion by the freeze concentration effect (see Figure 5.1), which in turn results in water being drawn from the plant's cells, partly dehydrating them and preventing them from freezing. This can occur since the membrane and wall of the plant cell prevent the ice outside the cell from seeding the freezing of the contents. Some plant tissues, such as the buds of leaves and flowers, enclose substantial spaces that lie outside the actual plant tissue. Ice forming in these spaces (extratissue or extraorgan freezing) results in water being withdrawn from the plant tissues themselves. Dehydration of the plant tissue occurs because the vapour pressure of the water within the tissue is higher than that of the adjacent ice. This partial dehydration prevents the plant tissue from freezing. This process occurs in the overwintering flower buds of rhododendrons, which are natives of cold mountain regions of North America and Asia – such as the Himalayas, for example. It is somewhat similar to the 'protective dehydration mechanism' of cold hardiness that is found in earthworm cocoons and some springtails. The large mass of plants means that they take a long time to freeze, and a slow rate of freezing is an important part of their ability to survive.

The seeds, and some other tissues, of many plants have very low water contents in their dormant state. They may be anhydrobiotic or the little water that is present cannot freeze. They can thus survive to very low temperatures since there is no freezable water. At low water contents, and particularly when there are high concentrations of sugars, cells may be able to survive very low temperatures by their intracellular water vitrifying. In this state, the water forms a glass-like solid without forming ice crystals.

The cold and frost tolerance of many plants changes with the season as a result of biochemical and physiological changes induced by the onset of winter. However, winter also represents a period of dormancy for the plant and it may be difficult to disentangle changes that are associated with dormancy from those directly responsible for the survival of low temperatures. The plant stores food to survive the winter and for the resumption of growth in the spring. Some of the food stores, such as sugars, may also be acting as cryoprotectants. Dormancy is triggered by changes in day length, as the days shorten during autumn, but low temperatures may also trigger cold hardiness more directly. The changes that occur during cold hardening are triggered by the production of abscisic acid. This plant hormone is involved in the response to other environmental stresses, such as desiccation.

Membrane lipids can solidify at low temperatures, disrupting the physiological function of the membrane. The temperature at which the membrane changes from a fluid to a solid or gel state depends on its lipid composition. Cold hardening of plants involves an increase in the proportion of unsaturated fatty acids in the membranes. Since unsaturated fatty acids are more fluid than saturated fatty acids, this enables their membranes to remain functional to much lower temperatures. Many plants accumulate sugars (particularly sucrose, but also glucose and fructose) and sugar alcohols (such as sorbitol and mannitol) during winter. These may act as cryoprotectants, as they do in animals, but they also act as food stores. Trehalose is not generally found in plants, but sucrose plays a similar role by binding to membranes and proteins, protecting them against dehydration.

A number of proteins are synthesised in response to low temperatures by cold-hardy plants. Some of these have been shown to play a role in preventing freezing damage. Several of the proteins induced by low temperatures are related to the dehydrins, produced in response to desiccation stress, and to the late embryogenesis-abundant proteins of seeds (see Chapter 3). This is perhaps not surprising since a major stress resulting from extracellular freezing is the dehydration of cells and their membranes. Some of the proteins that are induced by cold stress have been shown to have cryoprotective effects in assays which test their ability to preserve membrane function during freezing. The production of molecular chaperones, or heat shock proteins, in response to cold may be involved in stabilising proteins at low temperatures.

Antifreeze proteins have been isolated from carrots and winter rye. They appear to control the size and shape of ice crystals forming in the plants, preventing them from damaging cells. They also affect the stability of the ice by inhibiting recrystallisation. These proteins are located in the outer cell layers and intercellular spaces of the plants. They are thus likely to interact both with ice coming into contact with the surface of the plant and with that forming within its intercellular spaces. This suggests that they may play a role in modifying the growth of ice crystals in the plant.

MICROORGANISMS AND LOW TEMPERATURES

Since they are smaller and simpler than most plants or animals, microorganisms are more at the mercy of the conditions which surround them. When the temperature of their environment falls, they are directly exposed to the stresses associated with the cold and freezing. Microbiologists distinguish between microorganisms which are cold adapted, with an optimum growth temperature of 15 °C or lower (psychrophiles), and those which are cold tolerant, with normal optimum growth temperatures (20–40 °C) but tolerating low temperatures and even growing slowly at 0 °C (psychrotolerant). There are, of course, many more psychrotolerant than psychrophilic microorganisms since the metabolic machinery and structural components of the latter need to be adapted to work at low temperatures. Psychrophilic organisms, however, can only grow when there is liquid water present and growth will cease once their habitat freezes. Soil water is likely to freeze just below 0 °C and seawater freezes at about −1.9 °C. Unfrozen pockets may remain, however, and microorganisms are able to grow within these to much lower temperatures. Although microorganisms can not grow when the water surrounding them freezes, many survive and can resume growth when the water melts again.

Many different types of psychrotolerant microorganisms can be isolated from warm temperate soils, but psychrophiles tend to be absent since they can not compete with microbes which can grow better than them at warmer temperatures. There are, however, many environments that are frequently or permanently cold, favouring the presence of psychrophiles (see Chapter 2). Nearly three-quarters of the Earth is covered by deep oceans. Microorganisms living in deep ocean sediments experience conditions that are permanently cold (1–3 °C), as well as at high pressure (see Chapter 6). Almost constantly cold conditions are also present in polar regions and at high altitudes, associated with glaciers and permanent snow. In less extreme situations, the organisms may be exposed to low temperatures on a daily or seasonal basis with a corresponding change in the balance between conditions favouring the growth of psychrophiles and those favouring the growth of non-psychrophiles. Despite the cold conditions, however, Antarctic soil and aquatic habitats contain more psychrotolerant than psychrophilic microorganisms. It may be easier to tolerate the cold and wait for warmer conditions than to become adapted to growing at low temperatures.

A wide variety of microorganisms are found in cold environments, including bacteria, archaea, fungi (and yeasts), single-celled algae and protists. They are found in soil, the sea, lakes, streams and associated with plants and animals. Psychrophilic microorganisms have a number of adaptations that enable them to function at low temperatures. An increased proportion of unsaturated fatty acids, and other changes in lipid composition, enables their membranes to remain fluid and retain their physiological function. The enzymes of psychrophiles work best at low temperatures. This appears to be a result of changes in their structure that makes them more flexible in the cold, allowing them to continue to catalyse biological reactions. The structural proteins of psychrophiles, such as those that comprise the microtubule scaffolding of cells in eukaryotes (tubulin), are also stable at low temperatures.

A sudden decrease in temperature (cold shock) or continual growth at low temperatures (cold acclimation) stimulates the production of specific proteins. The cold shock response involves the production of stress proteins, in a similar fashion to the heat shock proteins produced in response to exposure to high temperatures (see Chapter 4). Cold shock proteins may play a similar role in removing cold-damaged proteins and acting as molecular chaperones, which assist the correct formation of other proteins in cells. The function of proteins produced in response to cold acclimation is less clear but they may play some cryoprotective role.

As well as coping with the cold itself, microorganisms may have to tolerate the freezing of their surroundings. This is likely to occur in terrestrial polar habitats, temperate soils during winter, and in sea ice, snow and glaciers. Microorganisms even persist in permanently frozen soils (permafrost) and they have been isolated from Siberian permafrost from a depth of 400–900 metres, the soil of which dates from the second half of the Pliocene (3–5 million years ago). The microorganisms found in permafrost were living in the soil, or were blown there by the wind, before it froze and reflect the climate of the region when conditions were more temperate. This explains why more psychrotolerant than psychrophilic microorganisms can be isolated from permafrost. They have survived for such a long time in a state of cryptobiosis.

Since microorganisms are mainly single celled, there can be no ice formed within them unless they freeze intracellularly. There are a few reports of microorganisms surviving intracellular freezing, but it is generally thought that the cell wall and plasma membrane prevent external ice from seeding their freezing. The formation of ice in the soil, or other medium, surrounding them will raise the concentration of salts, creating an osmotic gradient that dehydrates the cells. It is this dehydration that is the major stress for microorganisms during the freezing of their surroundings. Their survival is thus assisted by a slow rate of freezing, allowing them to adjust to the resulting dehydration. The thermal inertia of a large bulk of soil, or even rock, means that it is likely to take a long time to freeze. Some microbes produce sheaths or coats of mucus (extracellular polysaccharides), which may prevent immediate contact between their cells and ice in their environment. Antarctic yeasts and algae accumulate polyols and sugars which may be acting as cryoprotectants, while some bacteria accumulate amino acids in response to osmotic stress. These may play a role in freezing tolerance. Proteins with antifreeze activity have been isolated from bacteria.

In the early 1970s, it was discovered that some bacteria associated with the surface of decaying vegetation have strong ice-nucleating activity, seeding ice formation at temperatures as high as $-1\,°C$. These bacteria have proteins associated with their outer membranes which act as a template for the formation of ice crystals. A variety of ice-nucleating bacteria, particularly from the genus *Pseudomonas*, have been isolated. There are also some ice-nucleating fungi (*Fusarium*) and ice-nucleating activity is associated with the fungal component of some lichen symbioses. What advantage ice-nucleating activity confers on these microorganisms has been a matter of speculation. Since they are plant pathogens, seeding ice formation may damage the surface of the plant and enable the organisms to invade it. Nucleating activity may encourage the condensation of water onto the surface of bacteria that have been carried into the atmosphere and thus help them return to the earth in rain. The ability of these organisms to survive freezing may be assisted by ensuring that the plant tissues with which they are associated freeze at a high subzero temperature.

By encouraging the freezing of plants, ice-nucleating microorganisms are responsible for substantial amounts of frost-related crop damage in freezing susceptible plants. Conversely, they may aid the survival of freezing-tolerant plants by ensuring freezing at high subzero temperatures and thus preventing intracellular freezing. Both ice-nucleating bacteria and fungi have been isolated from the guts of insects and freezing-tolerant frogs. For a freezing-tolerant animal, the presence of these microorganisms may assist survival by producing freezing at a high subzero temperature, but, for a freeze-avoiding animal, they would be harmful since they would prevent supercooling. This may be part of the reason why some freeze-avoiding insects empty their guts during winter.

RESISTANCE AND CAPACITY ADAPTATION TO LOW TEMPERATURES

In environments that are temporarily cold, most organisms show resistance adaptation to subzero temperatures, surviving in a dormant state until conditions favourable for activity and growth return. Plants do not generally grow at low temperatures and most terrestrial animals, and other organisms, enter a state of dormancy or cold stupor and become inactive. Metabolism may slow sufficiently at very low temperatures for the organism to become cryptobiotic. Some freeze-avoiding insects, however, can remain active at very low temperatures. Mass aggregations of springtails are often observed moving across the surface of the snow in the European Alps ('snow fleas') even during the coldest part of the winter. The enzymes and physiological systems of these animals must be able to function at these low temperatures and they thus show capacity adaptation, as do psychrophilic microorganisms that can grow at low temperatures.

Organisms that live in environments that are permanently cold must maintain activity and growth at low temperatures. The seas of the Arctic and around the Antarctic are constantly at subzero temperatures. The fish, invertebrates and other organisms that live there must have capacity adaptation to the conditions.

COOL APPLICATIONS

Studies of the responses of organisms to cold have yielded, or have the potential to produce, a wide range of practical applications. These include the control of pests by influencing or predicting their survival overwinter and improved methods for cryopreserving organisms and biological materials. Cold-adapted enzymes from psychrophilic microorganisms and polar fish may have applications in the food industry, in biotechnology, laundry detergents and in the treatment of wastewater. Cold-adapted microbes themselves are used for the cold fermentation of beer and wine and for the ripening of cheeses and other foods.

Ice nucleation-active microorganisms have been used experimentally in food processing where freezing at a high subzero temperature is required (such as the freeze concentration of soy sauce, coffee, non-heated jams and other foods). As well as improving the quality of food processed by freeze concentration or freeze drying, they can also improve the texture of frozen foods, presumably by influencing the size and shape of ice crystals. Ice-active bacteria (*Pseudomonas syringae*) are widely used commercially for the artificial production of snow by ski fields (there is even a company in subtropical southern Georgia,

USA, which will supply the locals with an otherwise unlikely 'white Christmas'). The bacteria are sold under the trade name 'SNOMAX©' to be added to the water used by snowmaking machines. This assists snowmaking by raising the critical temperature for snow formation by as much as 20 °C. A similar product (SNOMAX weather manager©) is sold for seeding clouds to encourage rainfall or snow. It may be possible to control freeze-avoiding insect pests by exposing them to ice-nucleating microorganisms that reduce their capacity to supercool. On the other hand, it may be possible to prevent frost injury to plants by controlling their associated populations of ice-nucleating microorganisms.

A wide variety of ice-active proteins are produced by cold-adapted organisms, which influence the formation and stability of ice. These include ice-nucleating proteins which produce ice formation at high subzero temperatures and antifreeze proteins that inhibit ice nucleation. Antifreeze proteins produce a thermal hysteresis (a difference between the melting and freezing point in the presence of an ice crystal), but they also have the property of inhibiting recrystallisation. It is becoming clear, however, that some organisms (plants, molluscs and nematodes) produce proteins that inhibit recrystallisation but show little or no thermal hysteresis and cannot, therefore, be called antifreeze proteins. These may represent a third class of ice-active proteins, whose functions are to control the size, shape, location or stability of ice rather than the temperature at which it forms. Understanding how organisms control the properties, formation and stability of ice could find uses in a wide range of situations.

For the production of a creamy delicious ice cream, it is necessary for the freezing process to produce millions of tiny ice crystals, rather than fewer larger crystals which would give the ice cream a gritty texture. Ice-active proteins may be useful in controlling the size and nature of the crystals formed in ice cream and other frozen products. They may also be useful for inhibiting the ice recrystallisation that might cause the quality of the product to deteriorate. The nature of ice crystals formed in foods stored frozen also determines its quality when it is consumed after thawing and/or cooking. The addition of fish antifreeze proteins to meat before freezing results in the formation of smaller ice crystals. A number of studies have looked at whether the addition of fish antifreeze protein could improve the cryopreservation of sperm, oocytes, liver tissue and a variety of other cells and tissues. These studies have met with mixed success but companies have been formed to develop and exploit the commercial potential of these proteins. Attempts to improve the frost resistance of plants by inducing them to express fish antifreeze proteins via genetic engineering have so far been unsuccessful. The recognition that some plants themselves produce ice-active proteins may help in the development of techniques to improve frost resistance in plants. Ice slurries, mixtures of ice and water, are used for cold storage and heat transportation. The formation of large ice crystals by recrystallisation could block the pipes through which the slurry flows. Antifreeze proteins are able to prevent this.

Our understanding of how some organisms are able to survive freezing conditions is increasing and, although the ability to freeze human bodies is still some way off, it is clear that the lessons we learn from these organisms are going to find some cool applications.

6   More tough choices

So far, we have looked at organisms that can survive extreme temperatures (hot and cold) and desiccation (anhydrobiosis). There are, however, a number of other extreme environmental stresses that offer challenges and opportunities to living organisms. I will consider some of these in this chapter.

UNDER PRESSURE

There are few situations where organisms are naturally exposed to low pressure, but high pressure is a rather more common hazard than we might expect. Organisms that inhabit rocks and sediments beneath the surface of the Earth are likely to be under pressure (see Chapter 2 in the section 'The underworld'). The study of these organisms is in its infancy. We know rather more, however, about those of the other main high-pressure environment, the deep sea (see Chapter 2, 'The cold deep sea'). The deep sea is considered to be that volume of the oceans which is below the depth of 1000 metres. The oceans cover 71 per cent of the Earth's surface and are an average of 3800 metres deep. In volume, the deep sea comprises 75 per cent of the biosphere. This makes it the largest environment, or rather group of environments, on Earth, but it is one of the least understood. Since hydrostatic pressure increases by 1 atmosphere for every 10 metres in depth, organisms inhabiting the deep sea have to cope with the crushing forces of very high pressures – reaching up to 1100 atmospheres in its deepest parts. However, pressure is not the only problem organisms face. The deep sea is constantly cold (2–3 °C), dark and nutrients are in short supply since it is a long way from the primary productivity of the phytoplankton in the surface waters. Deep-sea organisms have to rely on nutrients drifting down to them. In some places, though, heat and/or food are available thanks to geothermal activity or the seepage of methane (see Chapter 2, 'Hot vents and cold seeps'). Pressure is, however, a pervasive feature of the deep sea. How do the organisms which live there cope with its crushing forces?

Pressure effects

Deep-sea organisms are not crushed by the high pressure because the pressure within their bodies is the same as it is outside, but they still need to be adapted to their high-pressure environment. Biochemical reactions are accompanied by changes in volume. If a reaction results in an increase in volume, it will be inhibited by pressure, whereas, if it is associated with a decrease in volume, it will be enhanced. Deep-sea organisms thus need to produce a different balance between their metabolic reactions than do organisms at the surface. Pressure also has effects on the structure of proteins and membranes. High pressure reduces the fluidity of membranes by squeezing their molecules together. Compression may also affect the conformation (folding) of proteins, so that it reduces the efficiency of their biological functions or stops them altogether. Ultimately, very high pressures may denature proteins.

Most animals that inhabit shallow depths are rapidly disrupted by the effects of high pressure and die. Microorganisms, being simpler, are more tolerant and may remain viable if they sink to the depths from their usual shallow habitats. Conversely, organisms that are adapted to the deep sea may not survive if they are exposed to the lower pressures at the surface and some animals will even fall to pieces as they are raised from the bottom. The ability of gases to dissolve in water is increased at high pressure. This causes problems if the pressure decreases, as gases come out of solution. This results in decompression sickness (the bends) in divers if they surface too quickly. Some deep-sea bacteria contain gas-filled spaces (vacuoles) that will expand and literally blow the organism apart if they are brought to the surface. These effects have made the study of deep-sea organisms difficult, requiring the development of devices which maintain the pressure experienced at depth as they are lifted to the surface. Marine biologists use PRATs

(Pressure-Retaining Animal Traps), and other devices, to maintain the physiological condition of deep-sea organisms during their journey to the surface for study in the laboratory.

Deep-sea organisms

Two main deep-sea habitats are recognised. Pelagic organisms live in the water column, while benthic organisms live on the sea floor or just beneath it. Of the 25 000 species of fish in the world, only about 15 per cent inhabit the deep sea. This is a relatively low diversity of fish, given the extent of the habitat, which may reflect the low availability of nutrients and the other problems of living there. Just about all major groups of fish have species which inhabit the deep sea, although they tend to be dominated by groups that appeared early in the evolution of fish. Special adaptations are required to meet the demands of this environment. Many different groups of invertebrates are also found in deep-sea habitats and there is a tremendous diversity of benthic invertebrates (see Chapter 2, 'The cold deep sea').

The identification of microorganisms from the deep sea is complicated by that fact that many surface-dwelling microbes will stay viable, but dormant, if they sink to the bottom of the ocean. Any sample of deep-sea sediment will thus contain representatives of organisms from the whole water column above it, as well as those that are specifically adapted to live there. Microbiologists distinguish between organisms that are piezotolerant (also called barotolerant) that survive high pressures and those which are piezophilic (barophilic) that grow best at high pressures. Piezotolerant bacteria will not grow above 500 atmospheres and grow best at 1 atmosphere, whereas the optimum growth of piezophilic bacteria occurs at pressures above 400 atmospheres. One of the most piezophilic bacteria isolated grows best at pressures of 700–800 atmospheres. It continues to grow at up to 1035 atmospheres but will not grow at all at pressures below 350 atmospheres (Figure 6.1). The DEEPSTAR group from Japan's Marine Science and Technology centre has isolated several strains of bacteria from the Mariana Trench and other deep ocean sites. These include

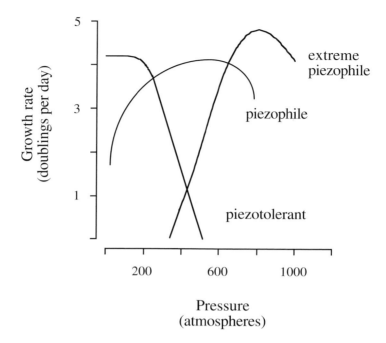

FIGURE 6.1 The effect of pressure on the growth rate of piezotolerant, piezophilic and extremely piezophilic bacteria. The extremely piezophilic bacterium was isolated from the Mariana Trench and will not grow at low pressures. Redrawn from a figure in Madigan *et al.* (2000).

strains that only grow at extreme high pressures (such as *Shewanella benthica*). Microorganisms are associated not just with deep-sea pelagic and benthic habitats but also with the guts, and other parts of the bodies, of the animals which live there. Hyperthermophilic archaea associated with deep-sea hydrothermal vents also live under high pressure. They are piezotolerant rather than piezophilic, growing more rapidly at 1 atmosphere than they do at the pressures experienced in their normal habitat.

Adaptations to high pressure
Some piezophilic microorganisms and deep-sea fish maintain the fluidity of their cell membranes by increasing the proportion of

unsaturated fatty acids in the lipid component of the membrane. The structure of proteins, enzymes and nucleic acids must also be adapted so they can function at high pressure. The enzymes of deep-sea fish seem to be insensitive to pressure, rather than being adapted to function best at high pressure. Proteins which are associated with cell membranes are particularly sensitive to the effects of pressure. Different outer membrane proteins are produced by the moderately piezophilic bacterium *Photobacterium profundum* at high and at low pressures. The switch in production from one outer membrane protein to another is controlled by the expression of a number of pressure-regulated genes. These proteins are thought to be involved in nutrient uptake, and the outer membrane protein produced at high pressures allows a faster rate of nutrient uptake in the low-nutrient environment of the deep sea, as well as being able to maintain its function at high pressure.

Fish produce trimethylamine oxide as a nitrogenous waste product (together with ammonia or urea). Cartilaginous fish (elasmobranchs like sharks and rays) retain urea in their blood and tissues, where it assists them with their osmoregulatory problems in seawater. Trimethylamine oxide is also retained and counteracts the harmful effects of urea on proteins. It is only found in low concentrations in shallow-water teleost (bony) fish other than elasmobranchs. Deep-sea teleosts and invertebrates, however, do contain high concentrations of this compound. It may, therefore, play a role in protecting these animals against the effects of pressure and has been shown to protect proteins against pressure-induced denaturation.

SWEET AND SOUR
The gardening column of my local paper tells me that I should be thinking about putting lime (calcium oxide) on my soil to sweeten it. Calcium oxide is an alkali (it produces a high pH in solution – see Chapter 1) and gardeners use it to counteract the acidity that tends to build up in the soil as salts are leached out by rain. Acids have a low pH and are sour. The sweetening effect of lime refers to it restoring the soil

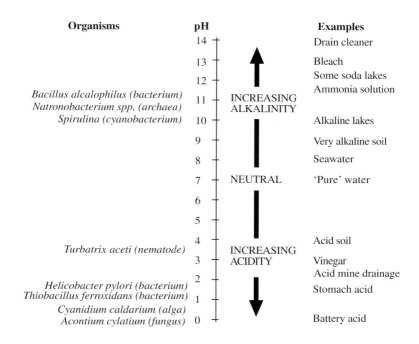

Organisms	pH		Examples
	14		Drain cleaner
	13		Bleach
	12		Some soda lakes
Bacillus alcalophilus (bacterium)	11	INCREASING ALKALINITY	Ammonia solution
Natronobacterium spp. (archaea)			
Spirulina (cyanobacterium)	10		Alkaline lakes
	9		Very alkaline soil
	8		Seawater
	7	NEUTRAL	'Pure' water
	6		
	5		
	4		Acid soil
Turbatrix aceti (nematode)		INCREASING ACIDITY	
	3		Vinegar
	2		Acid mine drainage
Helicobacter pylori (bacterium)			Stomach acid
Thiobacillus ferroxidans (bacterium)	1		
Cyanidium caldarium (alga)			
Acontium cylatium (fungus)	0		Battery acid

FIGURE 6.2   The pH scale in relation to some example solutions or environments and some organisms that can live at high pH (alkaliphiles) or low pH (acidophiles).

to a more neutral pH. Too much lime, however, is harmful and an alkaline soil can be as bad for plants as an acidic soil. Most plants grow best at a neutral pH (around pH 7). Some, however, grow naturally in soils that are slightly acid (peats and heaths) and should not be limed since the plants prefer to grow at a pH which is lower than neutral (these plants include rhododendrons, blueberries and huckleberries). Some soils are naturally alkaline (because they contain a high proportion of calcium and other salts) and the plants that grow in them tolerate a pH which is higher than neutral.

Most natural environments are at a pH of 5–9 (Figure 6.2). Very few organisms can grow in conditions that are very acid (below pH 3) or very alkaline (above pH 9). Those that do are called acidophiles or alkaliphiles. Those that can tolerate, but not grow, under these conditions are called acidotolerant or alkalitolerant. Growth at extreme pH is generally restricted to microorganisms. There is, however, at least one animal which can grow under very acidic conditions. Vinegar (about pH 2) can be made from any substance which contains sugar that can be fermented to alcohol and then to acetic acid, by the activity of yeasts and bacteria. It was probably first made from wine and the name means 'sour wine' (from the French). The vinegar eelworm (*Turbatrix aceti*, a nematode) was once a common inhabitant of vinegar, originating from the oak bark, twigs and branches that were used in its manufacture. Its natural environment was probably the fermenting sap issuing from the trees and other naturally fermenting habitats. This nematode was once so common in commercial vinegar that it was a widely held belief that its sharp taste was due the 'striking of these creatures upon the tongue and palate with their acute tails', a myth which was not discredited until the mid-eighteenth century. The vinegar eelworm can grow at an extraordinary pH range of 3.5–9, and can tolerate a range of 1.6–11, but it grows best at an acid pH. We have little idea how it can tolerate these conditions.

The production of vinegar is a result of the conversion of ethyl alcohol to acetic acid by the activity of acetic acid bacteria (*Acetobacter* and *Gluconobacter*). These bacteria are, of course, quite acid tolerant and can survive exposure to the acidity which they generate. A major use of vinegar, apart from flavouring, is for food preservation since few microorganisms can grow in the acidic conditions. There are, however, some acidophilic microbes which survive in some rather unusual environments that generate a very low pH.

Acidophilic bacteria, from a group called thiobacilli, are associated with hot springs, and other sites, which have high concentrations of sulphur. The thiobacilli use sulphur as an energy source, oxidising it to sulphur dioxide, which dissolves in water to make sulphurous acid. The bacteria oxidise this further to sulphuric acid. Their metabolic activities thus create an acid environment which they, and the other microbes that live there, can survive. *Thiobacillus ferroxidans* can use iron as an energy source and will grow in mine tailings that contain iron pyrites (iron disulphide). This bacterium can oxidise both the iron and the sulphur, producing acid in the process, and releasing some free sulphur. Other thiobacteria utilise this, adding to the acidity. A large pyrite dump can thus generate a very acid leachate, as a result of the activities of these microorganisms. This causes problems to mining engineers by corroding equipment and results in severe environmental problems. Several groups of archaea are acidophilic, including *Sulfolobus*, from sulphur-rich acid hot springs, and *Thermoplasma*, which is found in coal refuse piles. Hyperthermophilic archaea are associated with sulphur-emitting deep-sea hydrothermal vents. Some of these metabolise sulphur and are associated with acidic conditions. Other acidophilic microorganisms include fungi, yeasts, protozoa and algae.

The stomach of mammals is another common acidic habitat. Humans secrete 1.5–2 litres of gastric juices into their stomachs each day. The gastric juices are acidic (about pH 2) due to the presence of hydrochloric acid which is produced by the parietal cells that are found in glands on the stomach wall. Hydrochloric acid assists the initial stages of breaking down and digesting the food, but is also one of the first lines of defence against potentially harmful organisms. Any parasites which infect their host by being ingested along with food or drink must pass through the stomach before they can become established in their parasitic sites further down the intestine or elsewhere in the body. Many protozoan and animal parasites infect their host as a cyst or an egg. They are thus protected by a tough cyst wall or an eggshell and do not hatch or excyst until they have safely passed through the stomach. Their ability to survive chemical attack is remarkable and I have seen the eggs of some parasitic nematodes survive exposure to concentrated sulphuric acid.

There are few animal or protozoan parasites that live in the acid areas of the stomach. Until recently, it was assumed that there were no microbes that could do so either. The wall of the stomach is protected against attack by the acid and enzymes of the gastric juices by membranes and a layer of mucus. Occasionally, this protection breaks down, resulting in a peptic ulcer. For a long time, the reason for the formation of peptic ulcers was unclear and it was often put down to stress

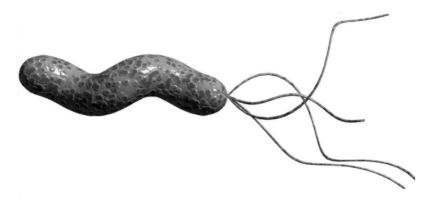

FIGURE 6.3 *Helicobacter pylori.* Drawing Luke Marshall, *Helicobacter* Foundation (from the *H. pylori* Research Laboratory website: www.hpylori.com.au).

or poor nutrition. In the early 1980s, two Australian researchers, Barry Marshall and J Robin Warren, presented evidence that peptic ulcers may be caused by infection with a bacterium, *Helicobacter pylori* (Figure 6.3). This suggestion was initially greeted with some scepticism since it was assumed that no bacterium could live in the acidic conditions of the stomach (although one has to wonder why, since the existence of acidophilic bacteria was well known by then). *H. pylori* is now recognised to be a major factor in peptic ulcers and other stomach problems. It is thought to be one of the commonest bacterial infections in humans, with 30 per cent of the population of the USA infected. About one-third of infected people develop peptic ulcer disease.

Naturally alkaline environments are even less common than acidic ones. They occur in soda lakes (see Chapter 2, 'Salt lakes and soda lakes') and in soils which contain high concentrations of carbonates. Alkaliphilic microorganisms (which have an optimum pH of 9 or higher) can, however, be isolated from a wide range of habitats such as soil and faeces, where they may be exploiting temporary pockets of alkaline conditions. Some archaea are alkaliphilic (in addition to being halophilic, see later) and inhabit soda lakes.

Although acidophilic and alkaliphilic microorganisms are very different, they face some similar problems. Both acids and alkalis are very corrosive, dissolving and destroying most biological materials. Parts of cells which are in direct contact with the external medium (surface membranes and flagella) must be able to resist these destructive effects. These organisms have to cope with either a surfeit or a dearth of hydrogen ions in their environment. Acids have a high concentration of hydrogen ions (low pH) while alkalis have a low concentration (high pH). The optimum pH for the activity of enzymes is around neutral (pH 7). Acidophiles and alkaliphiles would not be able to function if they allowed the pH inside their cells to reach that of their surroundings. They control their internal pH by regulating the movement of hydrogen ions across their surfaces. They can thus maintain an internal pH that is much closer to neutral than the solution in which they live. Despite this, the intracellular pH may be several pH units above or below neutral. These organisms thus have enzymes which work best under acidic or alkaline conditions. *H. pylori* can survive the acid environment of the mammalian stomach since it uses the enzyme urease to convert urea, which is a common waste product of the metabolism of nitrogen compounds, into carbon dioxide and ammonia. The ammonia dissolves in water to make an alkaline solution which neutralises the acid from the stomach in the region around the bacterium.

PASS THE SALT

You may like plenty of salt on your food, even though you realise that too much salt is bad for you. In fact, for most organisms, too much salt is fatal. High salt concentrations distort the structure of proteins causing them to stick together so they can no longer remain in solution. This destroys their biological function. Organisms may exclude the extra salt because their membranes prevent the salts from entering their cells. The problem then becomes one of the acquisition and retention of water.

A high concentration of salts in its environment means that the organism is in danger of losing water by osmosis (see Chapter 1 for an explanation of osmosis). Under these circumstances, organisms also find it difficult to gain the water they need from the environment since this involves the movement of water against the osmotic gradient (if the osmotic concentration outside the cells is higher than that inside, the water will move out of the cells and not into them). Despite these difficulties, there are some organisms that can thrive in environments which have very high salt concentrations. Seawater contains about 3 per cent salts and, since it is such a widespread environment, this can hardly be considered to be extreme. The salt concentration of seawater is fairly stable, except where it is diluted by inputs of freshwater from rivers. Salt concentrations higher than that of seawater occur mainly in terrestrial sites where the salts have become concentrated by the evaporation of water.

In some countries, salt is manufactured commercially from seawater. The seawater is allowed to flow into ponds which are then isolated from the sea (salterns). A concentrated brine is produced by evaporation, impurities are removed and the salt crystallised. If you fly into San Francisco International Airport, you may notice the large salterns on the coast which are used for the harvesting of sea salt. Natural salterns develop in areas which are flooded periodically by the sea. Salt lakes are formed where the rivers and streams feeding them flow over soils and rocks which contain easily dissolved minerals. They have no outlet and so water evaporation can result in very high salt concentrations (see Chapter 2, 'Salt lakes and soda lakes'). If the water evaporates completely, salt flats and pans are formed. If salts derived from salt lakes or seawater become buried, they form deposits of rock salt.

There are few organisms that can grow where there are very high salt concentrations. Those that can tolerate such conditions but grow better at low salt concentrations are referred to as being halotolerant, while those that grow best at high concentrations are called halophilic. Those which grow in high osmotic concentrations caused by substances other than salts (such as sugars) are called osmophilic. There are very few halophilic organisms that grow at salt concentrations between 15–30 per cent. The Dead Sea, for example, has just one halophilic alga (*Dunaliella parva*) and a few species of halophilic archaea (halobacteria). There are no plants or animals that can tolerate the conditions of the Dead Sea, which has the highest concentration of salts of any salt lake in the world (about 30 per cent salts). Lakes and ponds with a lower concentration of salt (up to about 25 per cent salts) can support the growth of brine shrimps (*Artemia* species) and the larvae of brine flies (*Ephydra* species). Halobacteria can grow in saturated salt solutions (33 per cent sodium chloride) and may even become trapped and survive within salt crystals.

Halophilic organisms can function in environments with high salt concentrations because they accumulate substances within their cells which counterbalance the osmotic stress. The alga *Dunaliella* accumulates glycerol. Glycerol is an osmotically active substance (osmolyte) which takes up space in a solution, lowering the concentration of water and raising the internal osmotic pressure. This balances the concentration of water inside and outside the cells, preventing them from losing water by osmosis. Glycerol is used as an osmolyte by algae, yeasts, fungi and *Artemia*. Prokaryotes, such as bacteria, use a variety of sugars, sugar alcohols, amino acids and compounds derived from these (such as glycine betaine and ecotoine) as osmolytes. These substances are known as compatible solutes because they do not adversely affect the workings of cells.

For most organisms, salts are not compatible solutes since they have harmful effects on their cellular machinery at high concentrations. Some halobacteria, however, accumulate high concentrations of potassium chloride inside their cells, although they exclude sodium ions. This involves pumping ions across the membrane and the halobacteria gain the energy required to do so in a unique way. Their membranes contain light-sensitive purple pigments called rhodopsins. These convert light energy into chemical energy and, in the process, use it to power their ion pumps. The rhodopsins give halobacteria their colour and accumulations of salts in salterns and salt lakes are often stained pink by their presence. The enzymes, and other biological molecules, of the halobacteria need to be adapted to function in the presence of high concentrations of potassium. In fact, the cells of halobacteria cannot function without the high levels of potassium and the binding of sodium ions to their outer surface is essential for them to maintain the structure of their cell walls.

THE BREATH OF LIFE

You only need to stick your head beneath the bathwater for a minute or so to realise how important breathing is for us. We use the oxygen, which makes up about one-fifth of the air we breathe, to fuel our bodies by burning (oxidising) our food. We cannot do without oxygen for long, but there is a surprising variety of organisms that can and which inhabit environments where oxygen is in short supply or absent altogether. Habitats in which oxygen is at low concentrations (microaerobic) or absent (anaerobic or anoxic) are in fact quite common. They include: the muds and sediments of lakes, rivers, ponds and oceans; bogs, swamps, deep waters and waterlogged soils; compost heaps and sewage treatment plants; and the intestines of animals and some deep underground areas. These habitats are anoxic because they are remote from contact with the air or the oxygen they contained has been used up by the activities of organisms.

Anaerobic conditions prevailed during the early stages of the Earth's history. The oxygen in the atmosphere originated from the activities of photosynthetic organisms and only accumulated once the quantity of oxygen produced by photosynthesis exceeded the capacity of chemicals in soils and sediments to remove it (by being oxidised). Organisms which live in aerobic conditions (in the presence of oxygen) had to adapt to doing so, while strictly anaerobic organisms, which cannot survive exposure to oxygen, have persisted in those habitats where it is absent. Strictly anaerobic microorganisms cannot survive exposure to oxygen because it forms some highly toxic reactive ions and molecules. Among these are the superoxide ion (O_2^-), hydrogen peroxide (H_2O_2) and the hydroxyl radical (OH^-). Aerobic organisms have enzymes which destroy these toxic products of oxygen. These include catalase and peroxidase, which break down hydrogen peroxide, and superoxide dismutase, which destroys the superoxide ion. Microorganisms which are strictly anaerobic are killed by contact with oxygen since they lack these mechanisms for dealing with its toxic products.

There are many groups of strictly anaerobic bacteria, a few fungi and a few protozoa. Some of these use fermentation reactions to gain energy from their food. These reactions are, of course, anaerobic, but involve the first steps of the pathways used by aerobic organisms. Anaerobic organisms can only release part of the energy of their food. They only gain 1/18 of the energy released by an aerobic organism which can break down a sugar to water and carbon dioxide by oxidising it through the involvement of oxygen. It is only the fact that they are exploiting habitats where food, in the form of organic material, is plentiful (such as in a compost heap) that they are able to survive in spite of such inefficient food utilisation. They also have less competition for the food, since aerobic organisms are excluded. Since anaerobes do not break down their food completely, they excrete much more complex molecules than do aerobic organisms (which produce carbon dioxide and water as the end products of sugar metabolism). Different groups of microorganisms produce different end products including: ethanol, acetone, glycerol, lactic acid, butyric acid, hydrogen and methane. Some anaerobic bacteria use inorganic materials, such as sulphates, as their energy sources.

Growing plants are unlikely to be faced with anaerobic conditions since they produce oxygen as one of the products of photosynthesis. However, dormant stages (pollen, seeds and spores) can survive without oxygen for some time. Many animals can survive periods of exposure to anoxic conditions, at least in some parts of their bodies. We can survive such conditions ourselves. If we exercise vigorously, the blood cannot supply enough oxygen to the muscles for aerobic respiration and they begin to respire anaerobically. Lactic acid is produced as an end product which, if it accumulates, causes muscle fatigue and pain. Many invertebrates can survive periods of anoxia, switching to anaerobic metabolism and accumulating organic end products which, if oxygen returns, can be further metabolised.

There are a variety of groups of parasitic worms which live in the intestine of mammals and other vertebrates. These include nematodes (roundworms), cestodes (tapeworms), trematodes (flukes) and acanthocephalans (spiny-headed worms). The wall of their host's intestine is supplied with blood which carries oxygen, but, in the centre of the lumen, conditions are likely to be anoxic. The amount of oxygen available to a parasite will thus depend on its position within the intestine. These parasites have no respiratory or circulatory systems and rely on diffusion for the transport of respiratory gases and other substances. Some of the parasites are quite large. *Ascaris lumbricoides*, for example, which is a common nematode parasite of humans in developing countries, is up to 30 centimetres long. Even if there is some oxygen available to them, it is likely that the deeper tissues of the parasite will be anoxic because of their reliance on diffusion for the transport of oxygen. Although they possess the pathways which would enable them to metabolise sugars aerobically, they, in fact, utilise anaerobic pathways. Like anaerobic bacteria, they can afford to use these inefficient anaerobic pathways since they have plenty of food available to them in the intestine of their host.

Long-term survival of anoxia by animals is rare, but there are a few invertebrates which can achieve this feat. James Clegg from the University of California's Bodega Marine Laboratory has hatched the brine shrimp *Artemia* from cysts stored in anoxic conditions for four years (Figure 6.4). Their food reserves (trehalose, glycogen and glycerol) showed no decline during this period, indicating that they had ceased metabolising. Clegg estimates that he should have been able to detect a level of metabolism just 0.002 per cent of normal. The cysts thus survive in a state of cryptobiosis as a result of anoxia (anoxybiosis). This is a similar phenomenon to anhydrobiosis, which is a state of cryptobiosis induced by desiccation (see Chapter 3). It is all the more remarkable since metabolism ceases, even though the cysts are fully hydrated and are at room temperature. *Artemia* cysts will survive a number of severe environmental stresses including desiccation, osmotic stress, ultraviolet (UV) radiation and temperature extremes, as well as anoxia. A protective cyst wall, the accumulation of trehalose

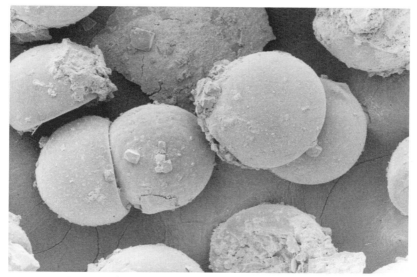

FIGURE 6.4 Scanning electron micrograph of cysts of the brine shrimp *Artemia*. The cysts are about 0.2 centimetres in diameter.

and glycerol, and the production of a heat-stress protein which acts as a molecular chaperone and of artemin (which is another protein specific to the cyst stage) may all be involved in its remarkable survival abilities. The survival of prolonged anoxia has also been reported in sponge gemmules (for four months), which are the protected dispersal stages of sponges, and in some nematodes (for 30 days). It may be much more widespread than we realise.

TOO MUCH SUN

I guess we are all aware that we should not spend too long in the sun, and that we must protect ourselves with clothing and sunscreens, if we do not want to end up looking like a boiled lobster. We are particularly aware of this in New Zealand, given our relatively low latitude (45° South) and proximity to the Ozone Hole which develops annually over Antarctica. Ozone levels over New Zealand have decreased by more than 5 per cent over the past 16 years. The sun gives the Earth life, via photosynthesis, but too much solar radiation is damaging to living

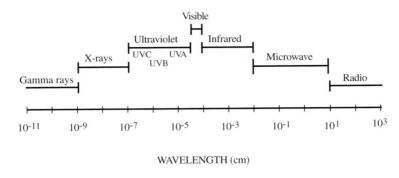

WAVELENGTH (cm)

FIGURE 6.5 The electromagnetic spectrum.

organisms and they need to cope with its harmful effects. The main destructive component is UV radiation. This is the part of the electromagnetic spectrum that has shorter wavelengths, and higher frequencies, than those of visible light (Figure 6.5). X-rays and gamma rays have shorter wavelengths still. The sun produces three classes of UV radiation: UVA (wavelengths of 400–315 nanometres), UVB (315–280 nanometres) and UVC (280–100 nanometres). UVC does not reach the surface of the Earth and UVB is the radiation that is responsible for most of the harmful effects on organisms, since it is the form that they most easily absorb. Not all radiation is bad, of course, and almost all life on Earth depends ultimately on the energy supplied by solar radiation in the visible range of the electromagnetic spectrum (400–700 nanometres).

The harmful effects of UV radiation (and other radiation, like X-rays) results from its high energy levels. When the radiation penetrates a living organism, it interacts with its molecules, resulting in damage and in the formation of ions and free radicals. These are highly reactive forms of atoms and molecules that will react with the molecules of cells to produce damaging effects. Of particular importance is the effect of radiation on chromosomes. The absorption of UV radiation by DNA damages genes, either directly or through their interactions with ions and free radicals, causing mutations. The damage may be repairable, but, if it is too great, there are disastrous consequences for the organism. Most mutations are harmful and may result in the cell dying or becoming cancerous. This process occurs, for example, in the development of skin melanomas as a result of too much exposure to the sun.

Earth's organisms are protected from most of the UV radiation from the sun since the atmosphere absorbs it. Ozone (O_3, one of the molecular forms of oxygen) is formed by the interaction between UV radiation and normal molecular oxygen (O_2). The absorption of solar radiation by O_3 and O_2 prevents almost all radiation with a wavelength of less than 290 nanometres from reaching the surface of the Earth – which is just as well, since if it did reach the Earth it would kill most of its organisms. This is why there is so much concern about the destruction of ozone in the atmosphere by chemicals which react with it (such as chlorofluorocarbons).

Despite the shielding effects of the atmosphere, significant amounts of UVA and UVB reach the surface of the Earth. Organisms may be sheltered from the sun under rocks, water or the surface of the soil. Hairs, feathers, scales, spines, skin and cuticle also provide some protection to the cells beneath. The skin itself is protected by melanin, a dark pigment which absorbs the radiation. The cuticles of insects, and other terrestrial invertebrates, contain dark pigments too and, in plants and some microorganisms, pigments may provide some protection. Many organisms have mechanisms that repair the damage done by the absorption of UV radiation. Radiation-induced breaks in DNA molecules are removed and DNA resynthesised by using the undamaged strand as a template (excision or dark repair). Some repair mechanisms depend on the presence of light, a process referred to as photoreactivation (light repair). The damage caused by UV radiation is thus lessened considerably if the organism is subsequently exposed to light in the visible range of the spectrum. In plants that are exposed to high levels of solar radiation – for example plants at high altitude – these repair mechanisms must be operating almost continuously.

Apart from solar radiation, organisms are exposed to other sources of radiation from terrestrial and extraterrestrial sources. Natural sources of terrestrial radiation derive from the decay of radioactive materials in the rocks, soil, air and water. In the past 50 years or so, there have been increasing sources of radiation from human-generated sources such as nuclear power plants and from radionucleotides used for medical diagnosis and therapy. Exposure to these radiations is normally low, except in some artificial situations. Non-solar extraterrestrial radiation consists mainly of cosmic rays which are high-energy particles originating outside our solar system. Any cells struck by one of these particles will be destroyed, but such collisions are rare. Damaged cells are easily replaced by plants and animals, and by the reproduction of surviving microorganisms.

Some microorganisms are able to tolerate high levels of radiation exposure. *Micrococcus radiodurans* is a radiation-resistant bacterium which was first isolated from a can of meat that had been exposed to high levels of gamma radiation to sterilise it. Another bacterium, *Pseudomonas radiodurans*, survives in the cooling waters of nuclear reactors. These organisms can survive high levels of radiation exposure because they have very efficient mechanisms for DNA repair.

THEY'RE TRYING TO POISON ME!

There are many substances that are toxic to organisms, but, in most cases, it depends on their levels of exposure. We can cope (quite happily!) with a moderate intake of alcohol (ethanol), but large quantities are likely to make us feel quite ill or even kill us. Even too much water can be fatal. Some substances, however, are toxic in low concentrations. Toxins are produced by some organisms themselves as protection against being eaten or to paralyse their prey. The arrow-poison frogs (*Dendrobates* and *Phyllobates*) of Central and South America, for example, have glands in their skin which produce a variety of toxins to discourage predators.

Mineral deposits can generate a toxic environment of non-biological origin. Natural accumulations of the element sulphur occur in association with volcanic activity and in sedimentary deposits. Some sulphur is essential for life; it forms part of the structure of many proteins. However, some of its compounds are extremely toxic. Toxic sulphur compounds associated with natural deposits of sulphur include hydrogen sulphide (the 'bad eggs' gas) and sulphuric acid. Most organisms do not tolerate these. There are, however, some bacteria which actually use the oxidation of sulphur compounds as an energy source (sulphur bacteria). *Thiobacillus*, for example, oxidises thiosulphate and elemental sulphur to sulphate. The bacteria that are associated with deep-sea hydrothermal vents, and which are in symbiotic associations with some of the animals that live there (see Chapter 2), gain their energy mainly from the metabolism of hydrogen sulphide.

High levels of heavy metals and some other elements (such as lead, copper, arsenic and antimony) are toxic to most organisms. There are microorganisms which can tolerate these elements and may even use them in their metabolism. These can be useful for cleaning up polluted areas. There are also microorganisms that can utilise any form of naturally occurring organic compounds which derive from biological activity. Microbes which degrade petroleum, and other types of hydrocarbon deposits, may be a nuisance under some circumstances, but are useful for cleaning up spills. Not all organic compounds are metabolised by microorganisms. Some of human origin (such as plastics, detergents and pesticides) are not degraded and thus accumulate in the environment, often with harmful effects.

COMPOUNDING PROBLEMS

Although this chapter, and the three which preceded it, have considered environmental stresses largely in isolation from each other, organisms are usually faced with a combination of physical and biological challenges. In Chapter 2, the stresses faced by organisms living in some of the extreme environments found on Earth were outlined. Desert organisms, for example, face problems of water availability, high temperatures, high solar radiation and a scarcity of nutrients. Terrestrial Antarctic organisms not only must survive low temperatures but also face problems with restricted and periodic access to liquid water. They may also be exposed to high levels of UV radiation during the summer and have to cope with transient and unstable substrates. There are common features between the effects of different environmental stresses on organisms and the response of organisms to them. Freezing, desiccation and osmotic stress, for example, all produce problems with the water content of cells. It is thus not surprising that the mechanisms which enable organisms to survive these three stresses have some features in common. I will look at this in more detail in Chapter 8.

7   'It's life, Jim, but not as we know it!'

As this quote from Mr Spock of the original *Star Trek* TV series highlights, science fiction writers (and scientists) have imagined many different forms of extraterrestrial life. These range from life consisting of 'pure energy', to life in the thermonuclear furnace of a star, to life within a few degrees of absolute zero which uses superconductivity to provide its energy. It may be rather hard to gain concrete evidence for the existence of such life, at least given our present abilities, even if such things do exist. Perhaps we would be better sticking to life as we know it. In this chapter, I will focus on what our knowledge of organisms in extreme environments might tell us about what is perhaps the greatest unanswered question in science: is there life elsewhere in the universe?

Organisms are found in a wide range of conditions on Earth. There are those that can live at temperatures above 100 °C, at high or low pH, or at high salinities, those that can survive temperatures below 0 °C, those that can dry out completely but survive, those that live at high pressures in the ocean's depths and those that can tolerate high levels of radiation exposure. The discovery of the communities of organisms associated with deep-sea hydrothermal vents in the 1970s (see Chapter 2) came as a complete surprise. Here was an ecosystem that did not derive its energy from sunlight but, instead, its primary producers, which produced the organic material, were chemoautotrophic bacteria that obtained their energy from chemical oxidations.

The discovery of extremophiles and cryptobiotes that can live in and/or survive a variety of extreme conditions, and the realisation that some utilise a range of chemical energy sources, means that we now know that life is possible under a much greater range of conditions than was ever dreamed possible. This realisation has helped fuel the present revival in interest in the possibility of life elsewhere in our solar system and beyond. Life on other planets is not only possible where there are 'normal' Earth-like conditions (moderate temperatures, pressures, pH and radiation exposure, and liquid water) but also where the prevailing conditions are similar to the extreme conditions that some organisms on Earth can survive. Life evolved under conditions on Earth that were more extreme than those existing today and hence may evolve under similar conditions on other planets.

WHAT IS LIFE?

Before we are able to discover life elsewhere in the universe, we must first know what we are looking for. There is no universally accepted definition of life, with ideas on the nature of life changing over the years. Different biologists favour different definitions. A physiological definition sees life as fulfilling a number of functions, such as feeding, growing, metabolising, excreting, reproducing, moving and responding to stimuli. We have already seen how the phenomenon of cryptobiosis challenges this physiological definition of life (see Chapter 3).

A metabolic definition considers an organism to have a distinct boundary (such as a membrane) which separates it from its non-living environment, but there is an exchange of materials with its surroundings, enabling the organism to maintain its structure by the consumption of energy. Again, cryptobiosis challenges this definition and it could also apply to a clearly non-living entity such as a candle flame.

A thermodynamic definition of life sees organisms as being in contradiction of the second law of thermodynamics. This states that the amount of disorder (entropy) is always increasing, with the universe (as a whole) moving towards a state of disorder and randomness. Organisms, at first sight, appear to defy this by maintaining order in the face of increasing disorder. Living things, however, are not closed systems. They maintain their structure by consuming energy from their surroundings – by absorbing sunlight, for example. There is thus a net increase in entropy (a decrease in order) as a result of the burning of the sun's fuel. Nevertheless, by absorbing energy, organisms maintain their structure, ultimately at the expense of an increase in the entropy of their surroundings.

Many biologists favour a definition of life which includes an evolutionary and a genetic or informational component. Living things include the instructions for their structure and functioning within genes which consist of nucleic acids (DNA and RNA) that can reproduce this information and pass it on to the organism's offspring. The replication of genes is, however, not always free of errors. These errors (mutations) are often fatal, but, occasionally, produce new structures or new metabolic processes which are advantageous to the organism. An organism that possesses such a favourable mutation is more likely to survive and reproduce than one which does not. Organisms are thus likely to accumulate favourable mutations and to become better adapted to survive and reproduce in their environment (or to become able to exploit new ones). This is, of course, the process of evolution and Darwinian evolution is seen by many as being central to our understanding of life. A recent definition of life which is frequently quoted was proposed by Gerald Joyce from the Scripps Research Institute in San Diego: 'Life is a self-sustaining chemical system capable of undergoing Darwinian evolution'.

WHAT IS NEEDED FOR LIFE?

Life (at least as we know it) needs a supply of the chemical elements which make up its composition, a source of energy and the presence of liquid water. All the organisms with which we are familiar also maintain a distinct structural integrity and are enclosed by some sort of a membrane that separates their cells from the environment, but which permits and controls the exchange of materials between the organism and its surroundings. Organisms also have nucleic acids (DNA and RNA) that encode the information that allows the production of the proteins and other molecules comprising their structure and functional systems. The nucleic acids also transmit this information to the organism's offspring during reproduction and changes in nucleic acids provide the raw material for the operation of the processes of evolution.

The main chemical elements that are involved in the structure of organisms are carbon, hydrogen, oxygen, nitrogen, phosphorus and sulphur. Sodium, potassium, magnesium, calcium and chlorine are also important as are, in trace amounts or in some organisms, manganese, iron, cobalt, copper, zinc, boron, aluminium, vanadium, molybdenum and iodine. These elements are widespread throughout the universe and their absence is unlikely to be a limiting factor for the development of life. During the eighteenth century, chemists developed the distinction between organic compounds which were isolated from organisms, and inorganic compounds which were derived from non-living material. The first artificial synthesis of an organic compound (urea) from inorganic compounds was achieved by the German chemist Frederich Wöhler in 1828. The development of life on a planet may require the presence of pre-existing (prebiotic) organic compounds. I will look at the potential sources of such compounds in the next section.

The most familiar source of energy exploited by organisms is sunlight, which is utilised by plants (and some other organisms) via the process of photosynthesis. As we have seen, however, some microorganisms can utilise chemical energy via the oxidation of various inorganic compounds. Geothermal energy and the energy derived from lightning and electrical discharges are other potential sources. Animals gain their energy by consuming plants (or microorganisms or other animals) while some microorganisms gain theirs by decomposing the bodies of other organisms. Sources of energy are likely to be widespread in the universe and will not be the main limiting factor for the development of life.

The requirement for liquid water is the most severe restriction for the development of life. Temperatures in the universe vary from those at the centre of stars (15 million degrees Kelvin [°K] at the core of our sun) to close to absolute zero (0 °K or −273 °C). Pure water is a liquid over a range of only 100 °C, a small fraction of the range of temperatures found in the universe. The range at which water is liquid may be extended slightly by mixing with other compounds (such as salts), which depress the freezing point, or under high pressures, which elevate the boiling point. Conditions which allow the presence of water as a liquid are, however, likely to be rare in the universe. One of the major roles of water in organisms is as a solvent and a medium in which biological reactions can take place in a controlled fashion. Could other liquids fulfil this role? Some other possibilities include ammonia (which is a liquid between $-33\,°C$ and $-78\,°C$), methane ($-164\,°C$ to $-182\,°C$), ethane ($-89\,°C$ to $-183\,°C$) and liquid nitrogen ($-196\,°C$ to $-210\,°C$). These are liquid at lower temperatures and over a narrower range of temperatures than water. Their lower temperatures mean that chemical (and biological) reactions would take place at a slower rate than they do in water and the fact that they are liquid at a narrower range of temperatures means that they are even less likely to be found as a liquid than water. They also lack some of the unique properties of water which are important in biological systems.

Water is a polar molecule, meaning that it has a slight negative charge at one end of the molecule (the oxygen end) and a slight positive charge at the other end (the hydrogen end). This polarity means that water molecules are attracted to each other and explains why many of the properties of water seem unusual in comparison with other liquids. The attraction between water molecules means that water is liquid at higher temperatures and for a greater range of temperatures than are comparable molecules. It also means that water has a higher surface tension and cohesion than do most other liquids and this helps organisms to absorb water from their environment and to transport it around their bodies. The ability of plants to absorb water through their roots and to transport it to their leaves via their water-conducting tissues (the xylem) depends on the cohesion of water.

Water has a high capacity to store heat and this has a buffering effect which ameliorates temperature extremes. The climate close to the sea is thus milder than that far inland. Water also absorbs a relatively large amount of heat when it turns into a vapour. This high heat of vaporisation, like the high heat capacity, tends to moderate the Earth's climate and thus improve its habitability to organisms. Terrestrial organisms can take advantage of the high heat of vaporisation of water to cool themselves through evaporative cooling.

The density of ice is lower than that of liquid water and that is why the ice in your gin and tonic floats. This prevents oceans and most lakes freezing solid at low temperatures since ice forming at the surface insulates the liquid water below. If the ice sank, all lakes and even the oceans would eventually freeze solid at low temperatures. Liquid water is so important for life that many consider that the search for life is equivalent to the search for liquid water.

THE ORIGIN OF LIFE ON EARTH

The Earth is thought to have formed 4.5–4.6 billion years ago, along with the other planets in our solar system, from a protoplanetary disc of gas and dust surrounding the sun (Figure 7.1). A period of intense bombardment from meteorites, asteroids and other protoplanets, swept up by the gravitational pull of the Earth, would have prevented the formation of life during the early stages. Life is thought to have arisen between 4.0 and 3.8 billion years ago. There is convincing fossil evidence for life 3.5 billion years ago and plausible evidence for life 3.85 billion years ago. This suggests that life developed on Earth as soon as there were conditions suitable for its existence. The major systems that needed to develop in order to produce something that we would recognise as being an organism are: a method of harvesting energy from some available source, the formation of the complex organic molecules found in organisms from inorganic or simple organic molecules, the development of an information-encoding and self-replicating molecule capable of supporting a process of evolution (such as DNA or RNA), and the formation of a membrane which encloses the organism.

Complex organic molecules may have formed on the Earth itself or have been carried to Earth from elsewhere. It is thought that the first organisms were heterotrophs, utilising these pre-existing organic compounds as a source of energy, rather than autotrophs, which produce organic compounds from inorganic compounds using the energy from

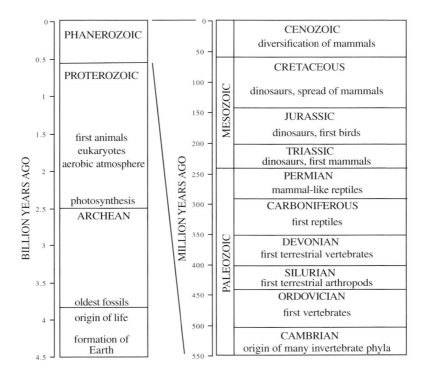

FIGURE 7.1 Geological timescale showing the major geological periods and some of the major events during the history of life on Earth.

sunlight or the oxidation of chemicals. The first attempt to simulate the formation of organic compounds on the early Earth was made by Stanley Miller in 1953, then a graduate student working under the supervision of the chemist Harold Urey. The Miller–Urey experiment consisted of a sealed glass apparatus in which a reservoir of liquid water was heated in the presence of an 'atmosphere' consisting of ammonia, methane, water vapour and hydrogen gas (Figure 7.2). Energy was supplied to the gas phase via an electrical discharge to simulate the effects of lightning. After running for several weeks, the apparatus was found to contain a wide variety of complex organic molecules, including hydroxy acids, aliphatic acids, urea and amino acids. Amino acids are the building blocks of proteins and they comprised about 4 per cent of the organic compounds formed in the experiment. Other sources of

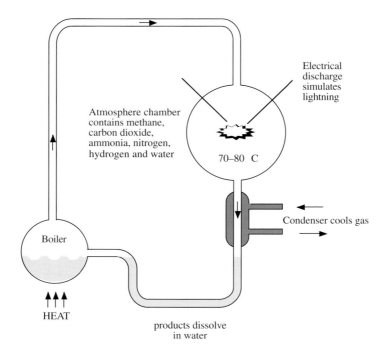

Atmosphere chamber
contains methane,
carbon dioxide,
ammonia, nitrogen,
hydrogen and water

Electrical
discharge
simulates
lightning

70–80 C

Condenser cools gas

Boiler

HEAT

products dissolve
in water

FIGURE 7.2 Diagrammatic representation of the Miller–Urey experiment.

energy, such as ultraviolet (UV) light, produce similar results. The production of organic compounds, however, is low if the conditions in the
vessel are not strongly reducing (high in hydrogen, low in oxygen). If
the carbon is present as carbon dioxide (rather than methane) and the
nitrogen as nitrogen gas (rather than ammonia), the yield of organic
compounds is reduced a 1000-fold. If the conditions on the early Earth
were strongly reducing, one can imagine the accumulation of organic
compounds in the oceans to form the famous 'primordial soup' from
which life could develop. In recent years, the most commonly held
view, however, is that the early atmosphere was not strongly reducing
but a weakly reducing mixture of mainly carbon dioxide, nitrogen gas
and water vapour. More strongly reducing conditions could still have
existed on a more localised basis, or for limited periods of time, allowing the accumulation of organic compounds.

The discovery of deep-sea hydrothermal vents in the late 1970s (see Chapter 2) suggested these as a potential site for the production of organic compounds. Water enters these systems from the sea by percolating through cracks formed as the Earth's crust cools and moves away from the mid-ocean spreading centres. It gets heated to high temperatures within these systems, which assists it to dissolve materials from the surrounding rocks. The water is injected back into the ocean at the vents at temperatures up to 370 °C, but rapidly cools as it mixes with ocean water. Several mechanisms for the production of organic compounds in these systems have been suggested and experiments that, it is claimed, simulate their conditions have produced organic compounds similar to those formed in the Miller–Urey experiments. Some, however, doubt whether these experiments simulate prebiotic conditions and point out that the high temperatures of vents are likely to destroy, and not create, organic compounds.

There are potential extraterrestrial sources of organic compounds. Comets, asteroids and meteorites impacted with the Earth at a high rate during its early history and material continues to do so today – fortunately at a lower rate. A wide variety of organic compounds have been detected in interstellar dust clouds by radioastronomy, which has detected nearly 100 different molecules in these clouds. These include the amino acid glycine. Organic compounds have been detected in comets and also in meteorites collected on Earth, including purines and pyrimidines which are components of nucleic acids. A class of meteorites called carbonaceous chondrites can contain more than 3 per cent carbon compounds with a variety of organic compounds, including amino acids and polycyclic aromatic hydrocarbons. It might be expected that organic molecules in meteorites, and other extraterrestrial bodies, would be destroyed by the heat generated during entry through the Earth's atmosphere. Micrometeorites (up to half a millimetre in diameter) and the dust from cometary tails are likely to have a soft landing, which would preserve their organic compounds. These are thought collectively to deliver 100 times more material to the Earth than do larger, but rarer, meteorites. They, together with the portion of larger bodies which survives cataclysmic impact with the Earth, could deliver significant quantities of organic compounds.

There are thus several potential sources of organic compounds for the early Earth. It is hard to judge their relative importance. If the atmosphere was strongly reducing, the sorts of mechanisms high-lighted by the Miller–Urey experiment were likely to be predominant. Under weakly reducing or oxidising conditions, hydrothermal systems and extraterrestrial sources may have been more important. Whatever the mechanism of their formation, these relatively simple organic compounds would need to interact to form the more complex com-pounds found in organisms.

Organic compounds dissolved in the seas of the early Earth were likely to form a fairly weak solution. In order to interact to form complex molecules, they needed to become more concentrated. One possibility is that they were absorbed onto the surface of clays or other minerals. Assuming the Earth was not completely covered by oceans, the margins of the land would have provided plenty of opportunities for the formation of pools or lagoons in which the organic material could become concentrated by the evaporation of water. Stanley Miller, who has maintained his interest in the question of the origin of life since his graduate studies in the 1950s, favours this 'prebiotic beach' hypothe-sis. The Moon and the Earth were much closer together early in their history and tides may have been 30 times greater than they are today. This would have allowed the formation of numerous tidal pools in which life could begin.

RNA is likely to have preceded DNA as the self-replicating informa-tional molecule of life, since it is a simpler molecule and has catalytic properties that may assist its functioning. In modern cells, RNA cata-lysts, called ribozymes, facilitate the formation of new RNA and a variety of reactions involved in protein synthesis. Since RNA can cata-lyse its own formation, it could have formed self-replicating molecules at a time when there was no DNA and no protein enzymes (referred to as the 'RNA world'). RNA is itself likely to have been preceded by simpler self-replicating organic molecules and/or self-replicating non-organic systems such as clays or minerals. Membranes may have developed from bubbles or froth formed in the primordial soup, by gas bubbles in hydrothermal systems or by the tendency of lipids, and other amphiphilic molecules (which have water-attracting and water-repelling parts), to assemble spontaneously into membrane-like structures. Such structures have been formed from amphiphilic compounds extracted from a meteorite. Membranes restrict the passage of materials through them and organisms have sophisticated transport systems which control the exchanges between cells and their surroundings. Primitive cells may have had membranes that were more permeable than those found today or that may have allowed the incorporation of materials from their environment during cycles of desiccation and rehydration.

As you will have gathered, there are plenty of competing theories of the origin of life among scientists (quite apart from religious interpretations!). Can our knowledge of the biology of organisms under extreme environmental conditions shed any light in this area? Conditions on the early Earth were likely to be much more extreme than they are today. Oxygen was lacking from the atmosphere and the first organisms would have needed to metabolise under anaerobic conditions. High concentrations of oxygen in the atmosphere only developed after the evolution of photosynthetic organisms, which were responsible for its production. Anaerobic bacteria are found today but are restricted to sites where oxygen is absent since they cannot deal with the potentially toxic products of an oxygen metabolism (see Chapter 6). Perhaps they are the survivors of the preoxygen world.

Temperatures on the early Earth may have been much hotter than they are today. The techniques of molecular systematics have given us an overall view of the tree of life (see Chapter 1). The most ancient groups of bacteria and archaea are hyperthermophilic. This suggests that the first organism (the last common ancestor) was also a hyperthermophile. This argues for a hot origin of life and for hot conditions on the early Earth. Of course, this is not accepted by some, who argue for a cool origin of life, as in the tidal pools envisaged by the 'prebiotic beach' hypothesis.

Anhydrobiosis may also be a primitive property of life. The primitive beach hypothesis envisages the concentration of materials in drying ponds and lagoons. These would presumably frequently dry out completely. Desiccation/rehydration cycles have been suggested to be important in allowing the entry of materials through primitive membranes. The involvement of anhydrobiosis in the origin of life is not a new idea. In the early 1960s, the late Howard Hinton (once Professor of Entomology at Bristol University) suggested that anhydrobiosis was a primitive property of cytoplasm. He thought that life arose on land rather than in the sea since not only would desiccation concentrate materials, but it occurred in numerous small ponds. This would allow more variation in both physical conditions and in the mixtures of chemicals contained within them than would be possible in the more uniform conditions of the sea. Such ponds could have provided an enormous number of experiments, just one of which could have resulted in the origin of life.

Life got started on the Earth once the period of heavy bombardment ceased. Even after this, occasional impacts from large asteroids, or other bodies, may have sterilised the Earth and required life to start again. Extremophiles can live in extreme conditions while cryptobiotes, in a desiccated state, can survive extreme environmental stresses that would be fatal to the fully hydrated organism. Extreme organisms would have had more chance of surviving large impacts and could have provided a remnant able to recolonise the Earth following such a cataclysmic event. Organisms may also have been protected at the bottom of the sea, around hydrothermal vents, deep within the Earth or under a covering of ice.

LIFE IN THE UNIVERSE

Earth is the only place in the universe where we know for sure that life exists. Those biologists who study life elsewhere (exobiologists or astrobiologists) have rather little to go on. The only other body in the universe that humans have visited, and from which samples have been returned to Earth, is our moon – which turned out to be a barren place as far as life is concerned. Astrobiologists have had to rely on understanding the conditions that have made life possible on Earth and then trying to determine whether such conditions exist elsewhere in our solar system and the universe in general. This is an area for endless, but entertaining, speculation and controversy. Whether extraterrestrial life exists is an important question, however, and proof of even very simple life elsewhere would have a tremendous impact on our understanding of our place in the universe.

The present revival in interest in astrobiology has been fuelled by our increased understanding of the range of conditions under which life exists on Earth, observations on the planets and their moons in our solar system made by space probes, and improvements in the instruments and techniques available to astronomers. This renewed interest has been recognised by the National Aeronautics and Space Aministration (NASA) by the establishment of its Astrobiology Institute. This is coordinated by the Ames Research Centre in Mountain View, California but is conceived of as a 'Virtual Institute' using the Internet to link a variety of researchers from around the world. The significance of extreme organisms has been recognised by NASA by its involvement in the 'Life in Extreme Environments' programme (LExEn) of the US National Science Foundation.

Until the 1960s, speculation on the possibility of life on other planets was confined to those based on earthbound observations. It is easy to forget how short the period of space exploration has been. The first satellite (Sputnik) was launched in 1957, the first manned space flight was in 1961 (by the Soviet Yuri Gagarin) and the first landing by humans on the Moon in 1969 (by the Americans Neil Armstrong and Edwin 'Buzz' Aldrin). The first probe to Venus was sent by the USA in 1962, followed by a series of Russian and American probes. Probes to Mars started in the late 1960s and have continued ever since. In the early 1980s, the US Voyager 1 and 2 probes sent back spectacular pictures of Jupiter and Saturn, and their moons, and of Uranus and Neptune. The pictures, maps and measurements taken by these space probes provide the basis for much of our present speculation concerning the possibilities for life in our solar system. We are visited by extraterrestrial objects (in the form of meteorites, if not UFOs!) and our ability to make observations from Earth (or near it) has improved with advances in the technology available to astronomers – particularly with the launch of the Hubble Space Telescope in 1990.

Our understanding of life, its possible origins on Earth and the range of conditions under which it can exist have increased enormously in the past 50 years. What does our knowledge of extreme organisms tell us about the possibilities for life in our solar system and elsewhere? I will just look at what are thought to be, or were once thought to be, the most likely candidates for extraterrestrial life.

Venus

Venus was once thought to be Earth's twin and people imagined a world of dense forests hiding below the atmosphere which obscures its surface. However, measurements by space probes have shown it to be an evil twin. Its atmosphere is about 100 times thicker than that of Earth and consists mainly of carbon dioxide. This has created a runaway greenhouse effect, trapping heat and producing average surface temperatures around 480 °C. Even at the poles or on high mountains, it is difficult to imagine life surviving such high temperatures which far exceed the capabilities of hyperthermophiles on Earth. Temperatures high in the atmosphere would be lower and some have suggested that organisms may exist which spend all their time floating in the clouds. Apart from staying permanently airborne, these would have to cope with very acidic conditions due to the sulphur dioxide which is injected into the atmosphere by volcanic activity at the surface.

Mars

Mars has long been our favourite place to imagine life elsewhere in our solar system and, despite us now recognising it to be a cold dry desert, it remains the most likely place to find evidence of past or present life. It is also the planet for which we presently have the most concrete evidence on which to base our speculations on the presence or absence of life.

In 1877, the Italian astronomer Giovanni Schiaparelli described what appeared to be straight lines on Mars, which he called 'canali' (channels). This was erroneously translated as 'canals' and led to intense speculation regarding their construction by a civilisation on Mars. The American astronomer Percival Lowell was a leading proponent of the idea that they were strips of vegetation bordering a system of irrigation canals designed to carry water from the poles. This no doubt inspired HG Wells to write the *War of the Worlds* (1898), in which he imagined the invasion of Earth by creatures from Mars, and in turn David Bowie's 1972 rock album *The Rise and Fall of Ziggy Stardust and the Spiders from Mars*. However, pictures of the surface of Mars from the Mariner 6 and 7 space probes in 1969 revealed no such canals, the appearance of which from Earth is now thought to be due to chance alignments of geological features.

There have been other apparent false alarms in our search for life on Mars. In 1975, Viking 1 and 2 released landers which arrived at the surface of Mars. The landers contained several instrument packages which sent information back to Earth, including biological experiments designed to look for life in the Martian soil. A scoop from the lander took a sample of soil which was distributed to the biology experiments. The first was a carbon assimilation experiment. Photosynthetic organisms on Earth absorb carbon dioxide from the atmosphere and process (assimilate) it into organic material using the energy of sunlight. Chemotrophic organisms can achieve comparable tricks using the energy from chemical oxidations. This experiment was designed to test for similar organisms on Mars. The Martian soil was exposed to carbon dioxide (and carbon monoxide) gas which had been brought from Earth and which was labelled with ^{14}C, the radioactive isotope of carbon. Any organic compounds formed from assimilation by Martian organisms would incorporate the ^{14}C and be labelled by it. The sample was then heated to break down any organic molecules, which would release the ^{14}C in the form of labelled carbon dioxide. This was measured by detectors which counted the radioactivity released by ^{14}C. You can imagine the excitement of the scientists when the first results of this experiment showed exactly what you would expect if the Martian soil contained active organisms. Heating the sample to 175 °C, however, reduced the activity, although it did not destroy it altogether. Destruction of any organisms by the heat should have stopped the assimilation of carbon. Even hyperthermophiles would not be expected to survive such high temperatures. It is, however, conceivable that cryptobiotes could survive if they were exposed to high temperatures in a desiccated state.

This first experiment was designed to look for autotrophic organisms which could fix carbon dioxide by using energy from non-organic sources. The second experiment was designed to look for heterotrophic organisms which use organic material as a source of raw materials and energy (usually by eating other organisms). The labelled-release experiment exposed the Martian soil to organic nutrients that had been labelled with radioactive isotopes. Any gases given off as a result of metabolic activity would also be labelled by these isotopes and could be detected. Again, the signal observed from this experiment was exactly what would be expected if organisms were present in the soil. Heating the sample to 160 °C destroyed the activity, which is also what you would expect if it was due to organisms.

The third experiment was a gas exchange experiment which detected the production of oxygen after the soil was exposed to a solution of nutrients. However, oxygen continued to be produced in the dark and after heating the sample to 145 °C – which is not what you would expect if the production of oxygen was due to biological activity (plants need light and any organisms would be destroyed by the heat). The final relevant experiment was a gas chromatography/mass spectrometry package which should have been capable of detecting any organic compounds in the soil. This failed to detect any. This equipment is so sensitive that the scientists concluded that not only was there no organic material present but that there must be some mechanism present which actively destroys organic compounds – some input of organic compounds to Mars from meteorites and comets would be expected.

The results of the Viking experiments gave some mixed messages as far as the presence of life was concerned. The metabolic experiments gave the results expected if organisms were present but it is hard to accept that life was present given the failure to detect organic compounds. The results of some of the control experiments (heating the samples to destroy any organisms) are also hard to reconcile with the presence of life. The consensus interpretation of these Viking experiments is that the results are due to some unusual chemical properties of Martian soil and do not indicate the presence of life. There are, however, some who believe that the results are consistent with, and are best interpreted as indicating, the presence of life.

The picture that the various space probes have given us of Mars is of a cold, dry, desert world. The average surface temperature is $-53\,°C$ but can rise above $0\,°C$ during daytime near the equator. The atmosphere is largely carbon dioxide and is thin, with about $1/100$ of the atmospheric pressure of that of Earth. The polar ice caps consist largely of solid carbon dioxide which melts and freezes on a seasonal basis. Beneath the carbon dioxide is a permanent ice cap of water ice. Water is present elsewhere on Mars, both as water vapour in the atmosphere and (probably) as ice beneath the ground. Liquid water is, however, likely to be rare and would rapidly evaporate into the atmosphere if exposed at the surface. There remains, however, the possibility of liquid water associated with hydrothermal systems, particularly below ground. The thin atmosphere of Mars gives little shielding from the sun's rays and UV radiation is intense at the surface. These conditions do not suggest the presence of life. However, our knowledge of extreme organisms shows us that extremophiles can exist under some very unpromising conditions and cryptobiotes can survive severe environmental stresses to exploit brief intervals of favourable conditions. NASA has recently (June 2000) announced evidence from the Mars Global Surveyor's Mars Orbiter Camera for the presence of recent (and perhaps even current) water flows on Mars, which, if confirmed, again raises the possibility of life (Figure 7.3). These features look like gullies and are formed on

FIGURE 7.3   Small gullies on the walls of the Nirgal Vallis, an ancient valley system
on Mars. Mars Orbital Camera images from the Mars Global Surveyor show more
than 14 of these channels, which are about 1 kilometre long and run down the
south-facing slope of the Nirgal Vallis. The lack of impact craters suggests that the
channels are relatively young (anything from a few million years old to just a few
weeks or days). They are thought to have been formed by groundwater seepage.
This is the only site where these types of features have been observed closer to the
equator of Mars than 30° latitude. Other sites are closer to the poles. Photo:
NASA/JPL/Malin Space Science Systems.

FIGURE 7.4 The canyon of the Nanedi
Vallis, one of the ancient valley
systems on Mars thought to have been
formed by water flows. Photo:
NASA/JPL/Malin Space Science
Systems.

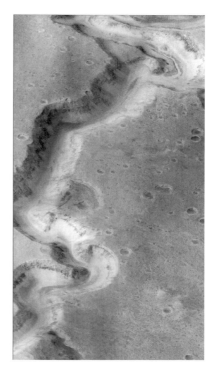

cliffs in craters or valley walls. It is suggested that they are formed by
water building up behind an ice dam, with the water being supplied by
seepage from a source beneath the surface. The ice dam eventually
breaks, sending a flood of water down the gully. The features are
thought to have formed relatively recently, and there may even be
active water flows today. The morphology of these structures is
similar to that formed by water flows on Earth. However, it is possible
that they are formed by some other liquid, such as liquid carbon
dioxide.

There is also evidence that conditions on Mars were not always as
inhospitable as they seem today and that it was once a warmer and
wetter world. There are flood planes, sedimentary rocks, river valleys
and erosional features that indicate the presence of abundant liquid
water at some stage in Mars' history (Figure 7.4). Conditions could

have once been much more conducive to life and its evolution. Maybe there are survivors that have adapted to the cold dry conditions of today. They could persist in rare favourable sites and, like cryptobiotic organisms on Earth, lie dormant waiting for conditions under which they can grow and reproduce. There may be fossil evidence of life from an earlier, more clement period. Some believe that such Martian fossils have already been found.

In 1996, a team headed by David McKay of NASA's Johnson Space Centre in Texas and Richard Zare of Stanford University announced that they had found evidence of fossil life in a Martian meteorite. The meteorite had been collected in Antarctica in 1984. Its reference number is ALH84001: ALH from its place of collection (Allan Hills, Antarctica), 84 is the year of collection (1984) and 001 the sample number. The Antarctic plateau is a good place to find meteorites since they are easy to spot in an otherwise featureless expanse of snow and ice. Ice flowing up against mountain ranges tends to concentrate the location of meteorites in this area. ALH84001 was originally classified as a diogenite meteorite – meaning that it was thought to be one of a group of meteorites that are considered to have originated from the asteroid belt. It was not until some 10 years later that it was recognised as belonging to that rare group of meteorites which originate from Mars. Of about 20 000 meteorites that have been collected, only 15 have been identified as being of Martian origin. They are thought to be the result of the impact of asteroids with Mars which ejected rocks from the planet into space and which have wandered the solar system before landing on Earth. It has been estimated that about half a ton of Martian material lands on Earth each year, but most of it, of course, goes uncollected. Scientists can identify these meteorites as coming from Mars because, in some, the composition of gases trapped within them matches that of the Martian atmosphere as measured by the Viking missions.

The evidence for fossil life in ALH84001 centres around the presence of visible deposits of carbonates, such as calcium carbonate (calcite). These minerals are formed by precipitation from water and

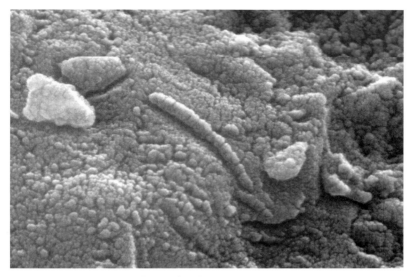

FIGURE 7.5 Possible fossil life found in the Martian meteorite ALH84001.
These structures are much smaller than bacteria, only 20–100 nanometres long
(a nanometre is one-billionth of a metre). Photo: NASA.

indicate the presence and involvement of water in the present structure of the meteorite. Of particular interest are carbonate globules
which are distributed throughout the meteorite. These show a distinct
layering around their edges, with alternating layers of magnesium-,
iron- and calcium-rich carbonates. On Earth, this type of layering is
produced where the precipitation of minerals is assisted by the activity
of bacteria. Small grains of magnetite (Fe_3O_4) and iron sulphide are
associated with the iron-rich carbonates. These minerals do not
usually occur together, except where the activities of certain bacteria
result in their simultaneous production. Organic compounds (polycyclic aromatic hydrocarbons) were detected in concentrations, and in a
distribution, that suggest they were not due to contamination from terrestrial sources.

Perhaps most intriguing was the presence of bacteria-like objects
associated with the carbonate globules which were observed by scanning electron microscopy (Figure 7.5). These are sausage shaped or look

like rice grains and have a very similar appearance to terrestrial bacteria. They are, however, much smaller, being about 100 times smaller than most bacteria. Terrestrial fossils of organisms in this size range (nanobes) have been reported, although there is controversy over whether these originated from living organisms. In March 1999, however, Philippa Uwins and a group of researchers from the University of Queensland announced the isolation of living nanobes from sandstone recovered from exploration wells. These look different from the proposed Martian fossils but are in the same size range.

Since the announcement of the evidence for possible fossil life in ALH84001, many non-biological explanations have been proposed, and evidence has accumulated which may contradict the biological origin of the structures observed. The controversy is unlikely to be resolved until we are able to send a mission to Mars to recover samples for analysis. Conditions in the Dry Valleys of Antarctica are the closest on Earth to those on Mars and are being used as a test ground for the development of techniques for searching for life on the red planet.

Jupiter and its moons

Jupiter is the most massive of the planets in our solar system. There are at least 16 satellites or moons orbiting Jupiter, the four largest of which (Io, Europa, Ganymede and Callisto) were discovered by Galileo in 1610. Jupiter is a gas giant with most of its volume being a thick atmosphere consisting mainly of hydrogen and helium. If there is a solid core to the planet, it is thought to comprise a relatively small proportion of its volume. The atmosphere of Jupiter is similar in overall composition to that of the sun. It generates heat not through nuclear reactions but by the continuing contraction of the planet under gravity as it cools. All the elements necessary for life are present in the atmosphere of Jupiter. Water vapour (which could exist as liquid water in the clouds) and some organic compounds (including methane, ethane, acetylene and hydrogen cyanide) have been detected. Conditions in the upper atmosphere are cool enough to allow life to exist. Deeper in the atmosphere, however, life would soon be destroyed by the heat and pressure. As with Venus, it is possible to imagine life forms floating in the upper atmosphere. It is difficult, however, to imagine how such life could have evolved.

At least one of the moons of Jupiter looks more promising. Some of the moons have a source of heat as a result of tidal heating due to their gravitational interactions with Jupiter and with each other. The orbit of Io around Jupiter is not exactly circular and so, at different times during its orbit, it comes closer to, or further away from, the planet. The gravitational pull of Jupiter on Io thus varies, distorting Io's shape. This gravitational distortion generates friction and thus heat. The tidal heating of Io is sufficient to keep most of its interior molten and there is abundant volcanic activity on the surface. There is no evidence for the presence of water (in any form) on Io and the existence of life is unlikely.

The amount of tidal heating on Europa is much smaller than on Io and its surface is cold. Remarkable images of the surface of Europa were acquired by the Voyager 1 and 2 spacecraft in 1979 and by the Galileo spacecraft in the late 1990s and the year 2000. From a distance, the surface is relatively smooth with few impact craters. This suggests that it has been regularly resurfaced. The absorption of sunlight from the surface of Europa indicates that it consists almost entirely of water ice. There are cracks and patterning, however, that indicate the presence of liquid water beneath the surface and the periodic flow of liquid water at the surface (Figure 7.6). It is possible that Europa is entirely covered by a frozen ocean of water and that there is liquid water beneath its surface. The subsurface water could be maintained as a liquid by the heat generated by tidal heating, from radioactive decay and from the penetration of sunlight. Estimates of the amount of heating generated from these sources vary widely and liquid water could exist at or just beneath the surface of Europa or the ocean could be entirely frozen.

Even if the ocean is almost entirely frozen, there could be pockets or lakes of liquid water maintained by frictional heating or volcanic activity. The discovery of Lake Vostok, and similar lakes beneath the central ice plateau of Antarctica (see Chapter 2), shows that such lakes can

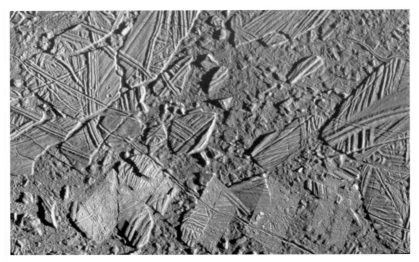

FIGURE 7.6 The Conamara Chaos region of Europa indicates relatively recent resurfacing. The irregularly shaped blocks of ice were formed by the movement and break-up of the existing crust. This is a mosaic of images taken by the Galileo spaceprobe in 1997. Photo: courtesy of NASA/JPL/Caltech.

exist even if covered by 4 kilometres of ice. Where there is liquid water, there could be life and Europa is the most promising place after Mars to look for extraterrestrial life in our solar system. Part of the interest in Lake Vostok stems from its similarity to the possible situation on Europa and its potential as a model to develop techniques to look for life under these conditions. The most probable source of energy for life on Europa is chemical oxidation with hydrothermal systems, driven by tidal heating, providing liquid water and an energy source.

Titan

Titan is the largest moon of Saturn and the second largest moon (after Ganymede) in the solar system. It is also the only moon in the solar system known to have a thick atmosphere. The atmosphere consists mainly of nitrogen and methane, together with hydrogen, carbon monoxide and carbon dioxide. Apart from methane, a variety of other organic compounds have been detected including ethane, propane, acetylene, ethylene, hydrogen cyanide, diacetylene, methylacetylene, cyanoacetylene and cyanogen. Water is thought to be present on Titan, but, with a surface temperature of $-179\,°C$, and no volcanic or hydro-thermal activity, the presence of liquid water is unlikely. Lakes or oceans consisting of liquid methane and ethane are possible. Life on Titan seems unlikely, but the presence of significant quantities of organic compounds suggests that it could mimic some of the processes that led to the development of life on Earth. A mission to Saturn, the Cassini spacecraft, was launched in 1997. In 2004, it will deploy the Huygens probe which will land on Titan and provide information on the composition of its surface and atmosphere.

Elsewhere in the universe

We know, of course, that life exists on Earth. Life on Mars seems possible, but not on Venus. There appears to be a habitable zone around the sun: with Venus being too close (too hot), Mars perhaps too far (too cold) and Earth just right. Apart from its distance from the sun, the size of a planet and its ability to form and retain an atmosphere will also determine its habitability. If planets occur around other stars in the universe, there could be earth-like planets within their habitable zones. Our understanding of the extreme conditions under which life is possible will affect our estimates of the sizes of habitable zones around stars.

The search for planets outside our solar system (extrasolar planets) has mainly relied on their effect on the motion of the star around which they are orbiting. Depending on how you define a planet, the presence of 53 extrasolar planets has been confirmed by July 2000. Using instruments such as the Hubble Space Telescope and the Very Large Array in New Mexico, protoplanetary discs have been observed around a number of young stars, including binary systems. From observations by the Hubble Space Telescope, it is estimated that there are 50 billion galaxies each containing up to thousands of billions of stars. Many of these may have planets around them, some of which will be within the star's habitable zone. Our solar system is unusual in that it involves a single star, whereas most stars are in binary or multiple systems.

However, given its size, it seems likely that there is life elsewhere in the universe. Whether we will ever be able to prove its existence is another matter.

LIFE IN SPACE: INTERPLANETARY TRANSFER AND THE REVIVAL OF PANSPERMIA

The observation that some anhydrobiotic organisms can survive exposure to the conditions of space (see Chapter 3) raises the possibility that they may be able to transfer between planets either naturally (in meteorites) or artificially (in spacecraft). Towards the end of the nineteenth century, the Swedish chemist Svante Arrhenius suggested that life arrived on Earth as microbes or spores wafting through space from elsewhere. This theory (called panspermia) merely transfers the problem of the origin of life on Earth to elsewhere in the solar system or universe. However, the realisation that material can transfer between planets in the form of meteorites does suggest a mechanism by which organisms could travel between them. Life originating on one planet could then seed other habitable planets in its solar system.

In 1969, the second manned mission to the Moon (Apollo 12) visited the site of Surveyor 3, a robotic probe which landed on the Moon in 1967. The astronauts recovered parts of the probe, including its camera, and returned them to Earth, under sterile conditions which would have prevented contamination. Scientists were astounded to discover viable colonies of bacteria in the foam insulating the camera's circuit boards. The bacteria were *Streptococcus mitis*, a common inhabitant of the mouth and throat of humans. Perhaps one of the technicians assembling Surveyor 3 sneezed and deposited these passengers which then survived the launch of the probe, its journey through space and more than two and a half years on the Moon. This is the only known survivor of unprotected space travel to another body in our solar system. The Apollo 12 Commander Pete Conrad realised the significance of this finding:

> I always thought the most significant thing that we ever found on the whole . . . Moon was that little bacteria who came back and lived and nobody ever said [anything] about it.

Only with our present knowledge of the survival abilities of cryptobiotes and extremophiles can we realise the significance of the bacteria's journey. It also provides us with a warning. We must be careful not to contaminate other planets accidentally with organisms from Earth or *vice versa*.

8   An extreme biology

Extreme environments affect the responses of organisms at many levels, from the structure of their proteins and membranes, to the ways in which they carry out their lives, their interactions with their environment and other organisms, and their life history, behaviour and evolution. While it would be an unhelpful truism to say that organisms are adapted to live in the places in which they live, there are some particular features of the biology of organisms in extreme environments. The extreme conditions which they face affect all aspects of their biology – theirs is an extreme biology which may give us some important insights into the nature of life in general. In this chapter, I will explore whether we can see any general patterns for life in extreme environments. But, first, let us revisit what we mean by an extreme organism.

WHAT IS EXTREME, REVISITED

In Chapter 1, I proposed a definition of extreme organisms by reference to the life boxes of organisms: the range of physical conditions under which they exist (the non-biological features of their ecological niche). I also proposed that we could recognise two groups of extreme organisms. Extremophiles have life boxes which are beyond the limits of those of the majority of organisms. Cryptobiotes have life boxes which are the same as, or substantially overlap, those of the majority of organisms, but, when the conditions of their environment deteriorate beyond those under which they can maintain their life processes (metabolism), rather than dying (as do most organisms), they cease metabolism and await the return of favourable conditions. We could thus recognise extreme environments as being those which have rather few species of organism (and those which are present are extremophiles) and/or those in which there are, at least periodically, species in a state of cryptobiosis.

I have mentioned a number of extreme environments in this book (places that are hot, cold, dry, salty, acidic etc.). I chose them, to be honest, not because they fitted the life box definition but because they seemed extreme to a human like me. In spite of this, most do in fact seem to fit the definition. There are fewer species that live in extreme environments (such as deserts and polar regions) than live in environments which are not extreme (such as tropical rainforests and coral reefs). There is, however, at least one extreme environment for which this is not true. We think of the deep sea as being an extreme environment because of the high pressures faced by the organisms that live there. Now that the problems of sampling organisms from this environment have been overcome, we have realised that, rather than being a biological desert, as had been assumed, it is populated by a very diverse range of species. The diversity of deep-sea organisms may be even greater than that found on land (see Chapter 2, The cold deep sea). The deep sea thus does not fit the definition of extreme that I proposed in Chapter 1. This definition focusses on the range of conditions that the majority of species could survive. Extreme environments would thus be expected to have a low diversity (rather few species would tolerate the physical conditions found there), whereas non-extreme environments would have a high (or, at least, a higher) diversity. According to this criterion, either the deep sea should not be considered to be an extreme environment or there is something wrong with the definition. I will return to this problem later.

EXTREME ENVIRONMENTS
Different extreme environments present different challenges to the organisms that inhabit them. Table 8.1 outlines the main extreme environments that were described in Chapter 2. These environments differ in the major physical and biological stresses they present to organisms, whether the stress is constant or varies and, if it varies, whether it is predictable. A periodic stress may have a degree of

Table 8.1 *Classification of extreme environments*

Extreme environment	Major stresses	Constancy of stress	Predictability of periodic stress	
			Long term	Short term
Terrestrial polar	Cold Desiccation	Periodic	Seasonal	Unpredictable
Marine polar	Cold	Constant	—	—
Saline lakes	Salt	Constant	—	—
Mountains	Temperatures, desiccation, ultraviolet radiation, oxygen	Periodic	Daily & seasonal	Unpredictable
Deserts	Heat Desiccation	Periodic	Daily & seasonal	Unpredictable
Temporary deserts	Desiccation	Periodic	Unpredictable	Unpredictable
Temperate mid-winter	Cold Food availability	Periodic	Seasonal	Unpredictable
Deep sea	Pressure	Constant	—	—
Hot springs Hydrothermal vents	Heat	Constant	—	—

predictability on a long-term basis since conditions are likely to be harsher during one season than another and harsher during the day than they are at night. Conditions may be unpredictable on a short-term basis, with temperatures and water availability changing on a daily, or even an hourly, basis. Where the stress is constant (as in polar seas, saline lakes, hot springs), the organisms must use capacity adaptation with their biological processes functioning under the conditions that are experienced in their environment. Where the degree of stress changes on a seasonal basis, organisms can respond by a period of dormancy (such as hibernation) and/or the production of compounds that aid survival of the stress (such as the seasonal production of antifreezes by some polar and temperate invertebrates). Periodic stress is likely to favour resistance adaptation, with the organism surviving the extreme conditions in a dormant state until conditions favourable for growth and reproduction return.

EXTREME ORGANISMS

We have met quite a few extreme organisms during the course of this book. They are found among all the major groups (Kingdoms or Domains) of organisms (plants, animals, fungi, protists, bacteria and archaea). However, within these groups, often only certain species, or groupings of species (phyla or lower taxonomic groupings), have the capacity to survive in extreme environments. Among the microbes, the Archaea are the organisms which can survive the most extreme environmental stresses. Some groups of bacteria, such as the cyanobacteria, are also prominent in extreme environments. Prokaryotes (bacteria and archaea) are generally more resistant than are eukaryotes (plants, animals, fungi and protists). In the Animal Kingdom, only certain phyla and species within phyla have solved the problems of living at extremes. Let us look at the distribution of extreme organisms within the Animal Kingdom to see if any patterns emerge.

Table 8.2 is a listing of some animals found in the most extreme environments and the stresses that they survive (based largely on examples from this book). While this is an incomplete list, some patterns are apparent. Only a few groups of animals live in extreme environments. Nematodes, arthropods and vertebrates (particularly mammals and birds) feature most prominently. This may simply reflect the fact that extreme terrestrial environments are colonised by

Table 8.2 Animals surviving in extreme environments

Stress	Groups	Examples
Desiccation (see Chapters 2 and 3)	Anhydrobiotic nematodes, rotifers, tardigrades, insect larvae and crustaceans	Various nematodes, rotifers and tardigrades *Polypedilium vanderplanki*, *Artemia*
	Non-anhydrobiotic invertebrates	Ants, termites, beetles, grasshoppers and snails
	Mammals, birds, reptiles and amphibians	Camels, kangaroo rat, ostrich, dune lizards and spadefoot toad
Heat (see Chapter 4)	Molluscs	*Sphicterochila boisseri* (desert snail)
	Insects	*Ocymyrmex barbiger* (desert ant)
	Nematodes and insect larvae from hot springs	Chironomid larvae
	Annelids from hydrothermal vents	*Alvinella pompejana* (the Pompeii worm)
	Anhydrobiotic animals	See above under Desiccation
Surviving or avoiding freezing (see Chapters 2 and 5)	Arthropods	Numerous species of mites, springtails and insects
	Nematodes and other invertebrates	Antarctic nematodes, potato-cyst nematode, earthworm cocoons, enchytraeids, intertidal molluscs and anhydrobiotic animals
	Polar fish	Notothenioids, Arctic cod
	Amphibians and reptiles	Hatchling painted turtles, wood frogs
	Mammals and birds	Polar bears, musk oxen, lemmings, penguins and ptarmigan
Pressure (see chapter 6)	Invertebrates	Many groups
	Fish	Many species
pH (see Chapter 6)	Nematodes	Vinegar eelworm
Salinity (see Chapter 6)	Crustaceans, insect larvae	*Artemia*, brine fly larvae

those groups of animals that are most successful at colonising terrestrial environments in general. Extreme aquatic habitats support a wider variety of animal groups, but even here only a few species have solved the problems of coping with high temperatures, low pH and high levels of salinity. The problems of coping with high pressures in the deep sea have been solved by a wide variety of invertebrates and fish, but, as I will argue below, perhaps we should not consider the deep sea to be extreme.

It seems that there are two broad groups of organisms in extreme environments. Organisms that are small and simple (bacteria, archaea, fungi, algae, protists, nematodes, rotifers and tardigrades) have limited abilities to control the conditions within them. Their cells are more likely to experience the physical conditions of the environment that surrounds them than are the cells of organisms that are larger and more complex. These simple organisms must rely on their membranes and enzymes being adapted to work in the extreme conditions, the perhaps limited regulatory abilities of their cells (capacity adaptations) and their ability to survive periods of stress by becoming dormant (resistance adaptations, such as anhydrobiosis).

The second group of organisms (birds, mammals and arthropods) are larger, more complex and appeared later during the evolution of life on Earth. Their complexity allows them to regulate their internal conditions so that they are more independent of the conditions in the external environment. Insects can retain water in a desiccating environment since they have a waxy cuticle and other ways of controlling water loss. Birds and mammals can regulate their internal temperatures, so that it is remarkably constant even in hot deserts or in cold Antarctic conditions. These abilities to regulate the internal environment in the face of extreme and changing external conditions (homeostasis) can be considered to be capacity adaptations to extreme environments. These animals also have behavioural and physiological mechanisms that enable them to avoid the most extreme conditions experienced in their habitats. Hibernation, other forms of dormancy, migration and restricting activity to the night in deserts are all ways of avoiding exposure to the most extreme conditions.

EXTREME PHYSIOLOGY

As you read the chapters that considered specific extreme environmental stresses (Chapters 3–6: desiccation, heat, cold, pH, osmotic stress etc.), you might have noticed that there were some similarities between the problems these caused to organisms. There are also some common physiological and biochemical solutions that organisms have developed to overcome these difficulties. Let us look at some of these problems and solutions.

Problems with water

Water is essential for the functioning of cells and any environmental stress that disrupts the cell's water balance is a serious problem for an organism. Exposure to desiccation is the most obvious cause of water loss, but water is also lost, or has the potential to be lost, during exposure to other types of environmental stress. Osmotic stress produces the movement of water, resulting in the loss of some water if the concentration of salts outside the cell is higher than that inside (since the concentration of water is lower outside than it is inside). Heat increases the rate of evaporation of organisms exposed to desiccation and this results in increased rates of water loss from the surface of terrestrial plants and animals. Mechanisms for cooling the organism (such as sweating, panting and transpiration) also produce increased rates of water loss. Water loss from cells is also a problem during freezing. Freezing of the liquid surrounding the cells raises its osmotic concentration. Salts are not incorporated into the ice and so, as water molecules join the growing ice crystals, the salts in the remaining unfrozen water become more concentrated. This results in an osmotic stress that draws the water out of cells (see Figure 5.1).

Problems with water may be reduced by restricting the rate at which it is lost and by measures that tend to retain it within the organism. The rate of water loss is reduced if the membranes of cells, or the skins or cuticles of plants or animals, have a low permeability to water. Many plants and insects, for example, have waxy cuticles and water loss is restricted to openings (stomata and spiracles) in their surface coverings. Water may be retained by animals by recovering it from the breath, urine and faeces before it is lost from the body. Desert organisms tend not to use the evaporation of water as a cooling mechanism, since they cannot withstand the loss of water that this involves. Instead, they either tolerate the high temperatures or they avoid them by retreating to refuges during the hottest parts of the day.

One way of reducing the problems with water caused by conditions which result in an osmotic stress is to accumulate substances within the organism or cell that reduce the difference between it and its surroundings. These substances are known as osmolytes since they take up space in a solution, thus reducing the osmotic concentration of water and balancing its levels inside and outside the cell. A variety of substances are used as compatible osmolytes (so called because they do not adversely affect the working of cells at the levels of concentrations to which they accumulate). Amino acids (e.g. proline, glycine, alanine and serine), substances derived from amino acids (e.g. glycine betaine and taurine), nitrogenous compounds (e.g. urea), sugars (e.g. trehalose, sucrose) and sugar alcohols (e.g. glycerol, sorbitol) are all used as compatible osmolytes by different organisms. Sodium, potassium and chloride ions are the most important solutes in cells. Changing the concentrations of these ions in response to osmotic stress gives a measure of control of water balance, but the extent to which this can be achieved is limited, since these ions are incompatible solutes and they are usually toxic to the cell in high concentrations.

Sugars, sugar alcohols and amino acids are also produced as antifreezes by freeze-avoiding insects and as cryoprotectants by some freezing-tolerant organisms (see Chapter 5). Although these substances play a number of roles in tolerating low temperatures, their ability to act as compatible osmolytes assists in dealing with the problems of water balance that result from exposure to subzero temperatures. Sugars are also produced by anhydrobiotic organisms in response to desiccation, protecting membranes and proteins against disruption as a result of water loss (See chapter 3). However, by raising the osmotic concentration of the body fluids, increased levels of osmolytes may also help slow down the rate of water loss from the organism. If the levels of osmolytes are high enough, the organism may even be able to absorb water from an otherwise desiccating environment.

Springtails (collembola) are common inhabitants of soil. They have very permeable cuticles which show little resistance to water loss. As the soil dries out, the springtail is at risk of losing water and of dying, since only a few springtails can survive anhydrobiotically. *Folsomia candida* is a common soil collembolan that can tolerate prolonged exposure to soils at 98.2 per cent relative humidity. The animal would be expected to lose water in soils below 100 per cent relative humidity. The springtail prevents this by producing high concentrations of glucose and myoinositol (a sugar alcohol). This raises its internal osmotic concentration sufficiently to arrest water loss to the environment. The springtail then starts to reabsorb water from the surrounding atmosphere. This enables it to remain active in the root zone of plants through a similar range of drought intensities tolerated by the plants themselves.

The accumulation of osmolytes is thus a common response to stress that involves problems with water balance. Trehalose, for example, is produced (by different organisms and under different circumstances) in response to desiccation, low temperatures, high temperatures and osmotic stress.

Problems with membranes
Maintaining the integrity of cell membranes is essential for the survival of an organism. The outer plasma membrane separates the cell from the environment and controls the exchange of materials between the inside and outside of the cell. Membrane systems within the cell (such as those of the endoplasmic reticulum, nucleus and mitochondria) play many important roles in the metabolism of cells. Cell membranes consist of lipids, proteins and carbohydrates. The lipid component, in

particular, is important for maintaining the structure of the membrane and is susceptible to damage by changes in physical and chemical conditions.

Changes in temperature may produce a change in the state of the lipids and hence in the structure and properties of membranes. Membrane lipids melt at high temperatures. The membranes of thermophilic bacteria have a higher proportion of saturated fatty acids than do the membranes of bacteria that live at lower temperatures. This change in lipid composition makes their membranes more stable at high temperatures. The membranes of hyperthermophilic archaea are stable at even higher temperatures; their membranes have a different structure from those of other organisms, consisting of a monolayer rather than a bilayer (see Figure 4.4). This is responsible for their stability at extreme temperatures. Membrane lipids can also undergo a change in physical state and hence a loss of biological function as they solidify at low temperatures. The membranes of cold-tolerant organisms tend to have a higher proportion of unsaturated fatty acids than do those that live in higher temperatures (the reverse of the situation in thermophilic bacteria). Part of the cold-hardening process in plants involves an increase in the ratio of unsaturated to saturated fatty acids in their membranes (see Chapter 5).

The integrity and function of membranes is also threatened by desiccation. The structure of membranes is maintained by interactions between its lipids and water. As water is removed, the structure of the membrane changes and this can result in the fatal loss of cell contents once water returns. Trehalose (or sucrose in plants) is thought to replace water during desiccation, preventing changes in membrane structure and enabling organisms to survive anhydrobiotically (see Chapter 3). In plants, and perhaps other organisms, proteins (dehydrins) may also be involved in stabilising membranes.

Problems with proteins
Both the structure and metabolism of cells depend on the functioning of proteins. In order to perform their roles in cells, proteins must

maintain their correct folding or shape (conformation). A number of stresses disrupt the weak chemical bonds that hold the shape of proteins together, causing them to unravel, or otherwise change their shape, and lose their ability to function correctly (they denature). Denatured enzymes can no longer catalyse biological reactions and denatured structural proteins can no longer fulfil their roles in an organism. A stress which results in the permanent denaturation of a significant proportion of an organism's proteins brings about its death. Protein denaturation can result from high temperatures, low temperatures, desiccation, high pressures, high salt concentrations and acidic or alkaline conditions. Organisms have ways of preventing, or ameliorating the effects of, protein denaturation.

Molecular chaperones are a group of proteins that assist in the folding of other proteins, preventing them from sticking to each other and enabling them to attain a structure that allows them to function. These were first described as heat shock proteins, since they are produced in response to a mild heat stress (see Chapter 4). They are, however, triggered by every type of environmental stress that has been studied and they also function in unstressed cells. Protecting the conformation and function of proteins is a universal and ancient problem. Molecular chaperones are found in all organisms and must have arisen early in the evolution of life. It might be expected that molecular chaperones play a greater role in the biology of organisms from high stress/high variability environments than in those from low stress/low variability environments. As well as these molecular chaperones, other substances, such as trehalose, are involved in stabilising proteins and preventing their denaturation during exposure to stress.

Proteins that function as enzymes work best under a particular set of physical and chemical conditions. Where environmental conditions are outside the normal range, the enzymes must either be modified to function under these extreme conditions or the internal environments of cells must be modified to make them less extreme than those that surround them. Enzymes from organisms in cold, hot and high pressure environments are adapted to work under these conditions. These extremozymes are of interest for practical applications that require enzymes which function under extreme conditions.

Problems with genes

Just like proteins, the nucleic acids that carry the genetic code (DNA) and those responsible for its translation (RNA) can be disrupted by exposure to various environmental stresses. Such changes are produced by, among other things, exposure to radiation, high temperatures, chemicals and oxygen radicals. Changes in DNA produce mutations that are usually harmful to the organism. Cells have a variety of mechanisms that limit or repair damage to nucleic acids and thus prevent the production of proteins that are malformed and useless. DNA molecules consist of two strands. If one of the strands is damaged, it will no longer pair up with the undamaged strand. The several different types of DNA repair mechanism rely on this lack of alignment between damaged and undamaged strands of the molecule. Enzymes called DNA repair nucleases recognise the altered portion of the molecule and separate it from adjacent unaltered regions on the same strand. A new copy of the damaged region is then made from the equivalent undamaged strand and inserted into the damaged strand to replace the altered region. These DNA repair mechanisms are particularly efficient in organisms that live in environments that are likely to induce high levels of DNA damage, such as in *Pseudomonas radiodurans*, a bacterium that survives in the cooling waters of nuclear reactors. As well as DNA repair, there are mechanisms that protect genes from damage and chromosomes contain proteins (histones) that bind to their DNA and help stabilise it.

Stress produces a greater rate of mutation, and of recombination (the process by which sexual reproduction produces new combinations of genes in the next generation by shuffling of genes from parents), which natural selection may act on to produce adaptation to stressful environments. Organisms exposed to stressful conditions thus tend to produce more variable offspring than those that are not exposed to stress. The increased variability may have a positive effect on the evolution of the organism in stressful conditions, since evolution depends on there being variability for the process of natural selection to have its effect. Stress may also, however, produce mutations that are harmful to the organism and the overall effects of stress are not necessarily beneficial in terms of the evolution of an organism. However, if some of the mutations induced by stress produce adaptations that help the organism survive the stressful conditions, these will be selected for and the organism will evolve to become better adapted to its environment.

A physiological definition of extreme?

Organisms that live in extreme environments have specific physiological and biochemical solutions to the problems which confront them. Perhaps we could recognise an environment as being extreme by the presence of such adaptations among the organisms that live there. There is, however, the danger of producing a circular argument – an extreme environment is considered extreme because the organisms that live there have adaptations to live in an extreme environment. All organisms have adaptations to the environment in which they live. An enzyme of a polar fish that works at $-1.9\,°C$, or of a hot-spring bacterium which functions at high temperatures, no more defines that environment as being an extreme environment than does the presence of an enzyme that functions best at $20\,°C$ in a tropical fish. Maybe we could recognise an adaptation to an extreme environment as being one that was not possessed by the organism's ancestors, which inhabited less extreme conditions. I will look for an evolutionary solution to the problem of defining extreme environments later in this chapter.

EXTREME ECOLOGY

The characteristics of extreme environments

Species richness (the number of species present) is one of the measures that ecologists use as a gauge of diversity and to describe and categorise communities of organisms. Is there any evidence that extreme environments have low species richness, and are diversity measures useful

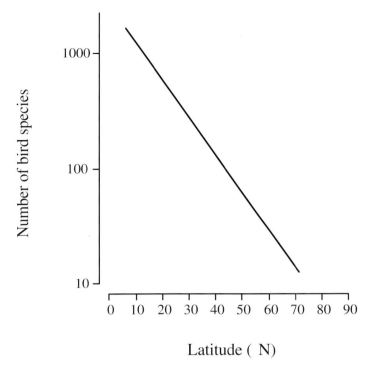

FIGURE 8.1 The number of species (species richness) of breeding birds in Central and North America declines as you go from the equator towards the North Pole (redrawn from Begon *et al.*, 1996, who give the source of the original data).

for identifying extreme environments? We could look for patterns in the species richness of a series of environments with an increasing challenge imposed by some physical factor. Average temperatures decline as you travel from the tropics to the poles; there is also a decline in species richness (Figure 8.1). This pattern of decline in species richness with increasing latitude is seen in many groups of terrestrial and marine organisms. It is difficult, however, to attribute it solely to decreasing average temperatures. Many other things change with latitude, including the range of temperatures experienced in any one place, the extent to which conditions change with the season, the availability of water in terrestrial habitats, the length of the growing season and the

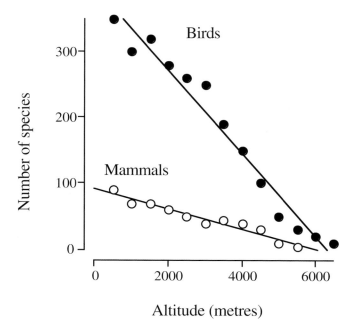

FIGURE 8.2 The number of species (species richness) of breeding birds and mammals declines with increasing altitude in the Nepalese Himalayas (redrawn from Begon *et al.*, 1996, who give the source of the original data).

amount of solar radiation available to fuel the growth of plants and other photosynthetic organisms.

Mountains provide an environmental gradient that occurs without a change in latitude. There are very few organisms that can survive the harsh conditions at the top of high mountains, but, at low altitude, depending on their location, there may be diverse communities of organisms. Several groups of organisms show a decline in species richness with increasing altitude (Figure 8.2). It is easy to accept that the environment becomes more harsh as you ascend a mountain, but the change is not a simple one and, in some ways, mirrors the changes seen with increasing latitude. Average temperatures decrease but temperature ranges increase with altitude. Organisms living at high altitude on tropical mountains experience sudden switches between very hot days

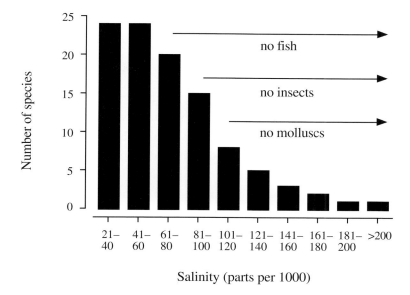

FIGURE 8.3 The number of animal species found in saline lakes of southeastern Australia declines with increasing salinity. This graph is based on the ranges of salinity tolerances of animals as given in Bayly & Williams (1973).

and freezing nights. Water and oxygen availability decreases with altitude, while winds and other sources of physical disturbance may increase. The area available for organisms to colonise decreases with increasing altitude and this could also be responsible for the observed decline in species richness.

Saline lakes vary in the concentrations of salts found dissolved in their waters. There are relatively few organisms that can live in highly saline lakes (those that have a salt concentration greater than 50 parts per 1000 or 5 per cent salt). The number of animal species found in the saline lakes of southeastern Australia declines with increasing salinity (Figure 8.3). For these lakes, there are no fish at salinities above 70 parts per 1000, no insects above 95 parts per 1000 and no molluscs above 112 parts per 1000. The most salt-tolerant animal in these lakes is a crustacean *Parartemia zietziana*, which has been found at salinities of up to 353 parts per 1000. There are no animals or higher plants living in the

Dead Sea, which has an average salinity of 370 parts per 1000, and it only supports one species of alga, a few species of halobacteria and some viruses. Thus, there seems to be a decline in the species richness of saline lakes with increasing salinity. However, the survival of organisms in saline lakes may be affected by factors in addition to salinity. Saline lakes may have high concentrations of substances other than sodium chloride. The Dead Sea, for example, has high levels of magnesium and calcium, which appear to be particularly difficult for organisms to cope with. High salinity also affects various physical characteristics of the water, including its density, freezing point and its ability to hold oxygen.

Productivity is another measure that ecologists use to categorise communities of organisms. This is the rate of production of new biomass (the total mass of organisms) and is related to factors such as the availability of light and nutrients, temperature and the length of the growing season. In general, we expect diversity to decrease with increasing harshness of the environment, but the pattern of change in productivity may be more complex (Figure 8.4). Dr Dev Niyogi, currently a postdoctoral fellow at the University of Otago, kindly provided me with Figure 8.4 and with some of the ideas on which the following interpretation is based. As environmental stress increases, diversity decreases, but productivity, as measured by biomass, initially increases. The reason for this is that, in a complex community of organisms, there are a number of levels in the food chain or food web. Plants use the energy of sunlight to produce organic material (they are the primary producers), the plants are eaten by herbivores and the herbivores are eaten by predators, which, in turn, may be eaten by bigger predators. Herbivores use the energy and organic material contained in the plants they eat to build their own bodies. This process is not efficient. The biomass of herbivores is always less than the biomass of the plants they have consumed. A community that consisted solely of plants (primary producers) would thus produce a greater total biomass than one which included herbivores and predators (all other things being equal). As the harshness of the environment increases, not only are there fewer species of organisms that can live in the conditions but

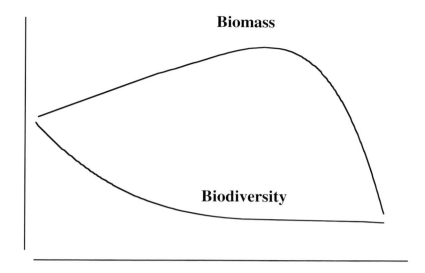

FIGURE 8.4 As the harshness of the environment increases, the biodiversity (e.g. number of species) decreases, but the biomass (total mass of organisms) initially increases before declining. Based on a drawing by Dr Dev Niyogi, University of Otago.

whole levels in the food chain disappear. Large predators are the first to go, followed by smaller predators and then herbivores. Ultimately, only the primary producers are left, plus perhaps some decomposers (such as bacteria) which rely on the dead bodies, or products of, the primary producers. The Dead Sea, for example, has just one primary producer (the alga *Dunaliella parva*) and several species of halobacteria that rely on its products. Finally, the conditions may become so harsh that no organisms can survive and the biomass declines to zero.

The decline in diversity and the increase in productivity in extreme environments suggests that the organisms which are able to colonise them gain a very great prize. The success of an organism, in terms of its numbers or biomass, depends not just on the physical conditions of its environment but also on its interactions with other organisms. It may

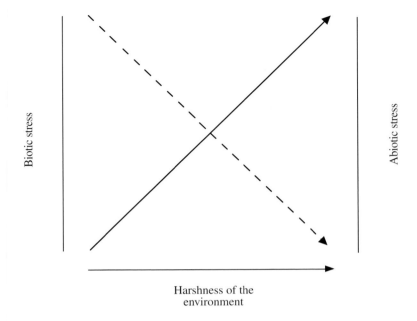

FIGURE 8.5 As the harshness of the environment increases, the amount of abiotic
stress (physical factors such as temperature, pH and, salinity) increases (——), but
biotic stress (predation, grazing, competition, parasites and diseases) decreases
(-----).

be eaten by other organisms, it may have parasites and diseases, and it
may face competition for resources from other organisms. The
increased productivity of extreme environments may be associated
with a reduction in these biological pressures (Figure 8.5). The alga *D.
parva* more or less has the Dead Sea to itself, with no other organisms
present that eat it or compete with it. An extreme organism may thus
be more successful (it can produce a greater biomass and/or occupy a
greater proportion of the available habitat) than one which lives in a
less extreme environment.

The first organism to colonise an extreme environment may be able
to exclude other organisms that seek to do so (a sitting tenant effect). If
an organism develops an adaptation which enables it to survive the
conditions, although the solution may not be an efficient one at first,

the process of natural selection may ensure that the adaptation becomes optimised. Any competitor which develops a similar adaptation will not have had the opportunity for it to become optimised and will be outcompeted by the sitting tenant. Perhaps the sitting tenant will only be evicted if the interloper comes up with a more efficient solution to the problem of coping with the conditions.

Although it seems to be generally true that species diversity is low in extreme environments, this might be driven by factors other than the harshness of the environment. Some extreme environments might be considered to be islands surrounded by less extreme conditions (as is the case with hot springs, hydrothermal vents and some saline lakes) or by even more extreme conditions (as in the ice-free areas of the terrestrial Antarctic that support moss and algal growth). These environments are small, rare and with a low complexity in terms of the sorts of habitats they provide within them. These sorts of island effects could in themselves be responsible for the low species diversity observed, rather than the extreme physical conditions *per se*.

The distribution and dispersal of extreme organisms

Extreme organisms are usually limited in their distribution by a requirement for the extreme conditions to which they have become adapted. Some may, however, be capable of colonising less extreme conditions but are prevented from doing so by competition from, or predation by, other organisms. Populations of the brine shrimp *Artemia*, for example, are limited to saline lakes and ponds that are too salty to support the fish, and other predators, which would eat them. The extreme conditions thus represent islands of habitat that are surrounded by a 'sea' of unsuitable habitat. Suitable habitats may be separated by quite large distances. Organisms need to be able to disperse and colonise new areas. If they stayed in the same place, they would be vulnerable to catastrophes that could wipe out their population and cause them to become extinct. How do extreme organisms cope with the necessity for, and problems of, dispersal?

The fauna of deep-sea hydrothermal vents illustrate the problem.

Hydrothermal vents are temporary, opening and closing as the tectonic forces that create them shift. The animals living around hydrothermal vents must colonise new vent habitats, if they are to survive catastrophes and leave descendants in the long term. New vents sometimes appear hundreds of kilometres away from any existing ones and yet are rapidly colonised by animals. Vent animals are specialised to live there and have to cope with the extremes and gradients of temperatures and the high concentrations of metals and sulphides they experience. They rely on sulphide oxidation by bacteria for their sources of organic materials, rather than the products of photosynthesis by algae – as do animals living in the rest of the sea. The vents are thus widely separated by an environment that will not support the growth of adult vent animals. Like most marine animals, the animals of hydrothermal vents disperse by producing larvae. These are not capable of actively swimming the distances involved and must rely on ocean currents to carry them from one vent site to another.

Terrestrial Antarctic organisms (such as mosses, nematodes, springtails and mites) are limited to areas of ice-free ground which receive sufficient moisture from melting snow and ice to support their growth. Suitable sites are rare and may be separated by large distances. Many of these organisms are capable of anhydrobiosis, at least in some stages of their life cycle. In a dry state, they can survive being blown around in the air and could be transported large distances by this method. Springtails have water-repellent cuticles and rafts of springtails have been observed floating on the surface of water. They could perhaps disperse along coasts by this method. Some Antarctic mites can survive prolonged immersion in seawater. Transfer on the legs of seabirds (e.g. skuas, gulls and terns) is another possibility.

Trade-offs by extreme organisms
Organisms vary in the way they reproduce and conduct their lives. This includes aspects of their life history such as their fecundity (how many offspring they produce), how long a period they reproduce for, their age before they start reproducing, how often they reproduce and the size of

their eggs (or other reproductive stages). The most successful pattern of reproduction might be to produce lots of large, well-provisioned eggs, frequently and early in the lifespan of the parent and to produce them for a long period of time. In practice, no organism can achieve such a strategy since food, and other resources, are limited. Organisms have thus had to evolve a compromise between the different characteristics of their life history. One trade-off is that parents who put a lot of resources into producing offspring might have to pay for this by having a shorter lifespan (or less growth) themselves. The resources they devote to their offspring cannot be assigned to supporting their own growth or survival. For example, a tree might produce more seeds or it might grow more quickly, but it cannot do both if it has limited resources. Another commonly recognised trade-off is between the number of eggs an organism produces and the size and amount of nutrients supplied by the parent to each egg.

Ecologists have proposed several schemes for classifying the life-history patterns observed in organisms. One commonly used scheme is to divide populations of organisms into 'r' and 'K' strategists (these refer to the characteristics of the growth curves of the organism's population). An r-selected population of organisms is thought to be adapted to short-lived or unpredictable environments by being able to reproduce quickly. These organisms are small in size, become reproductively mature early on, they may have a single large breeding event and produce lots of small offspring. A K-selected population of organisms is thought to be adapted to competing with other organisms in a stable environment. These organisms have a larger size, become reproductively mature later, may reproduce several times and produce fewer, but larger, offspring. For organisms in extreme environments, the patterns found in their life history depend on the predictability of the extreme conditions. An environment that is constantly and predictably extreme, would be expected to favour K-selected features, whereas an environment that is unpredictably extreme (or rather unpredictably favourable for growth) should favour r-selected features.

Organisms in extreme environments may have further characteristics that enable them to survive in harsh conditions. These are referred to as being *A/S* selected (adversity/stress selected). *A/S*-selected organisms take a long time to complete their life cycle, have low growth rates, low fecundity and rates of reproduction, a poor ability to compete with other species and an investment in survival strategies. Peter Convey of the British Antarctic Survey considers that terrestrial Antarctic plants and animals show the *A/S* selection pattern well. They have long life cycles, which are associated with low average temperatures, low water availability and short growing seasons. They also have physiological and biochemical mechanisms that enable them to survive low temperatures and desiccation.

In extreme environments, there may be three-way trade-offs between survival, growth and reproduction (Figure 8.6). Survival of low temperatures in terrestrial Antarctic organisms, for example, requires the production of sugars or polyols (trehalose, glycerol) and proteins (antifreeze proteins). Resources spent on producing these survival compounds cannot be spent on growth or reproduction. Where the severity of the environment varies with the season, there will be a seasonal shift in the production of survival compounds. Resources will be transferred entirely into survival during entry into winter and then shifted back into growth and then reproduction during the spring and summer (Figure 8.6A). Polar fish produce antifreeze proteins which enable them to survive the risk of freezing in polar waters. Since they are constantly exposed to this risk, the production of antifreeze proteins needs to be maintained. In a constantly extreme environment, the organism will achieve some sort of balance between the demands of survival, growth and reproduction (Figure 8.6B). This latter pattern may also be needed in environments where the stress is rapid and unpredictable, requiring the continual production of survival compounds to meet the risk. This might be the case in the more extreme terrestrial Antarctic habitats, or in rapidly drying habitats, where extreme conditions can occur at any time of the year and the change is too rapid to allow the production of protective compounds. Survival mechanisms thus have to operate continuously for the organism to persist.

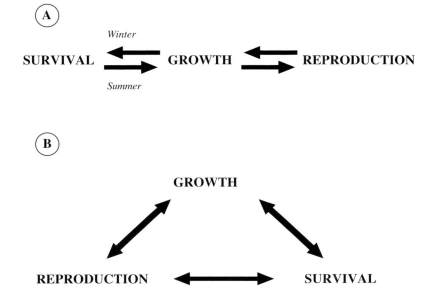

FIGURE 8.6 The allocation of resources to reproduction, growth and survival by organisms in extreme environments.

In (A), there is a seasonal shift in the allocation of resources as conditions become less extreme during summer and more favourable for growth and reproduction.

In (B), the conditions are constant and a balance between the resources allocated to survival, growth and reproduction is achieved.

EXTREME EVOLUTION

Like most things in biology, we can only truly understand the nature of extreme organisms and extreme environments in the context of their evolution. The conditions we observe today are but a snapshot in time and the organisms are a product of a long history of change, both in themselves and in the conditions which surround them. Conditions on the early Earth were very different from what they are today. It is in this context that we need to consider the nature of extreme organisms and environments. However, before I consider how an evolutionary perspective can shed light on our understanding of this area, let us look at one of the best studied examples of evolution in an extreme environment.

Frigid fish: a case study of evolution in an extreme environment
Notothenioid fishes dominate the fish fauna of Antarctic coastal waters. They make up 55 per cent of the species and 90 per cent of the individuals that are caught there. Their success in Antarctic waters is attributed to a single adaptation – the evolution of antifreeze glycoproteins. Although they face a number of problems in cold Antarctic waters, they could not persist there without a mechanism for preventing the ever-present risk of the freezing of their blood (see Chapter 5). They are thus a unique example of evolution in an extreme environment that can be attributed to the development of a single protein. After the fragmentation of Gondwanaland, Antarctica gradually drifted south, cooling as it did so. This represented a cooling of surface waters from about 20 °C down to the temperature seen today (-1.9 °C). Antarctica became isolated by the development of the Antarctic Circumpolar Current during the early Oligocene to early Miocene (38–25 million years ago – the Cenozoic Era, see Figure 7.1). This resulted in the cooling of Antarctic waters and antifreeze glycoproteins must have evolved by the mid-Miocene (about 17 million years ago). This is relatively recent in terms of evolutionary time, and means there is some chance of deciphering the evolutionary processes that occurred. Coastal fish living in Arctic and northern temperate waters are faced with similar problems and have developed similar antifreeze proteins.

Chi-Hing Cheng and Liangbiao Chen working in Professor Arthur DeVries's research group at the University of Illinois–Urbana have been unravelling the evolution of the genes responsible for producing antifreeze proteins. Antarctic notothenioid fish produce several different types of antifreeze glycoproteins (proteins with attached carbohydrates). Antifreeze glycoproteins are also produced by cod (Gadidae) from cold waters in the northern hemisphere. Other groups of fish produce antifreeze proteins (without the attached carbohydrates). Four different types of antifreeze proteins have been described. These are found in groups of fish that are only distantly related to one another or to the Antarctic notothenioids. This suggests that antifreeze proteins (or glycoproteins) have evolved independently among unrelated groups of fish several times. This is a striking example of convergent evolution – the development of a similar solution to an environmental stress by unrelated groups of organisms. A number of groups of fish living in coastal polar or northern temperate waters have independently evolved different types of antifreeze proteins and glycoproteins that protect them against the risk of freezing.

Cheng and Chen have determined the structure of the gene responsible for the production of the antifreeze glycoprotein of an Antarctic fish. They compared this gene structure with those of other genes, the sequences of which are stored in computer databases. The antifreeze glycoprotein turned out to be similar to one end of the gene sequence coding for trypsinogen from plaice (a flatfish). Trypsinogen is one of the steps in the production of the intestinal enzyme trypsin, which is a protease involved in the digestion of the proteins contained in the animal's food. This suggests that the fish antifreeze glycoprotein evolved from a gut enzyme, an example of an existing gene becoming modified to produce a protein with a new function. Ice crystals are continually being ingested by the fish along with food and water. The gut is thus a major site of potential ice nucleation and it is perhaps not surprising that a gut enzyme has become modified to deal with this threat. Cheng and Chen have now isolated a gene that represents an intermediate in the process; this produces both an antifreeze glycoprotein and a trypsinogen-like protease. Antifreeze glycoproteins that circulated in the blood, and other body compartments, must have developed from those which were secreted into the gut.

Genes contain sequences (exons) that code for proteins separated by non-coding sequences (introns, sometimes called 'junk' DNA since such sequences do not code for the production of proteins). The region of the trypsinogen gene that shows similarity to the antifreeze glycoprotein gene includes part of the non-coding sequence. Antifreeze glycoproteins thus owe their origin to 'junk' DNA and are examples of sense being created out of nonsense. This is a very rare demonstration of this method of evolving new genes. Using molecular clocks (changes in molecules whose rate of evolution is known), Cheng and her coworkers have determined that antifreeze glycoproteins evolved between 5 and 14 million years ago. This corresponds to the time that Antarctic waters cooled to freezing temperatures. By developing this molecule that enabled them to survive in Antarctic waters, notothenioid fish were able to fill the niches vacated by fish that were unable to survive the conditions created by the cooling of Antarctica.

The sequence of amino acids from the antifreeze glycoprotein of the Arctic cod (*Boreogadus saida*) is almost identical to that from an Antarctic notothenioid (*Dissosticus mawsoni*). These two Arctic and Antarctic groups are thought to have been isolated from each other for about 40 million years, long before the cooling of Antarctic waters and the evolution of the antifreeze gene. Differences in the structure and organisation of the antifreeze genes from the two fish reveal their separate evolutionary origins. No regions of the Arctic cod gene bear similarity to trypsinogen (unlike the Antarctic notothenioid antifreeze gene) and the gene appears to have had a different origin. Four different types of antifreeze proteins have been described in a number of groups of fish inhabiting polar and northern temperate waters, including some sculpins, sea ravens, eel pouts, wolf fish, smelt, herrings and winter flounder. The presence of antifreeze proteins in unrelated groups of fish indicates that the genes responsible for their production evolved independently. Antifreeze proteins are also found in terrestrial arthropods, plants and molluscs, and there is evidence of their presence in nematodes. They are found in situations where it is important for the survival of the organism to either prevent freezing or to control the size, shape or location of ice crystals in their body. The evolution of antifreeze (or ice-active) proteins thus represents a widespread response to the risk of freezing in environments where extreme cold is experienced.

What is extreme: an evolutionary solution?
The life box definition of an extreme organism, proposed in Chapter 1, considered that such an organism could survive environmental

conditions beyond those tolerated by the majority of organisms (see Figure 1.3). However, the conditions tolerated by organisms have changed. This is both because conditions on Earth have changed throughout its history and because the organisms themselves have evolved new abilities that enable them to live in places where they could not live before. Let us look at a few examples.

Life began in the sea and organisms had to solve a number of problems to move from an aquatic to a terrestrial environment. The land was probably first colonised by microorganisms. They would, however, have been restricted to habitats that were, at least periodically, wet. Microbes cannot grow unless they are immersed in water. Their ability to survive in an anhydrobiotic or dormant state would, however, have seen them through the dry periods. Plants colonised the land during the mid-Silurian period (420 million years ago, see Figure 7.1). They are thought to have evolved from green algae and needed to develop a waxy cuticle and stomata, to control desiccation, and roots to acquire water and nutrients and to anchor them to their substrate. Animals did not colonise the land until after the plants since they needed them for food and, perhaps, shelter. Arthropods (particularly spiders and insects) started to colonise the land about 410 million years ago and vertebrates (leading to the development of amphibians, reptiles, birds and mammals) about 374–360 million years ago.

Animals had to solve a number of problems to make the transition from a fully aquatic (living in water) to a fully terrestrial (living in air) lifestyle. Air is much less dense than water and so they needed some sort of skeleton to support their bodies. Terrestrial animals have to face much greater temperature extremes since they have forgone the thermal buffering properties of water. Marine animals can fulfil their requirements for water and salts from the seawater surrounding them, whereas terrestrial animals must acquire them by eating and drinking. Terrestrial animals must also maintain their water content in an often dry environment.

The problems of life on land were solved long ago and terrestrial organisms have since colonised every possible habitat on land. We do not think of most terrestrial habitats as being extreme, but, to an organism living in the early Silurian period, all terrestrial habitats would have seemed extreme (if it had been capable of thinking about it). At that time, the life box of the majority of organisms would have encompassed the conditions that were found in the sea, and perhaps some other aquatic habitats. The life box of a terrestrial organism would have lain outside the life box which encompassed those of the majority of organisms and hence the terrestrial organism would be considered to be an extreme organism, for its time. Terrestrial organisms have now become so numerous that the life box of the majority of organisms embraces most terrestrial conditions and they are no longer extremists. The recognition of extreme conditions and extreme organisms changes as conditions and organisms have changed.

As well as organisms changing to colonise an extreme environment (as in the movement from water to land), there have been dramatic changes in the conditions on Earth which have required organisms to change or become extinct. The early Earth had an anaerobic atmosphere, lacking in oxygen. The concentration of oxygen in the atmosphere only increased after the evolution of photosynthetic organisms, which produced oxygen as a product of their photosynthesis. Such organisms are thought to have evolved before 2.5 billion years ago (see Figure 7.1). Oxygen levels in the atmosphere would not have risen, however, until most oxidisable materials on the Earth had been exhausted. Iron and some iron minerals are unstable in the presence of oxygen (like a rusting nail, for example). The deposition of iron minerals in sedimentary rocks indicates that the absorption of oxygen by oxidisable minerals was complete by about 1.7 billion years ago. There was thus a fundamental change on Earth from anaerobic (no oxygen) to aerobic conditions (oxygen present in the atmosphere). Early organisms were anaerobes; to them, exposure to oxygen represented extreme conditions since they could not survive exposure to its toxic products. Organisms evolved that could cope with the aerobic conditions. Those that did not acquire this ability either had to survive in the few remaining anaerobic environments, or they became extinct. In the ancient anaerobic world, the aerobic organisms were the extreme organisms, but, now that aerobic conditions are prevalent, it is the anaerobic organisms that are the extremists. What once was extreme has become normal and what once was normal has become extreme.

Many researchers now conclude that, as well as being anaerobic, early organisms were thermophiles or even hyperthermophiles, living at much higher temperatures than are considered normal on Earth today. As the Earth cooled, organisms evolved which could survive and function in the cooler conditions. Hyperthermophiles are still with us today, but, at the Earth's surface, they are restricted to the relatively rare habitats of hot springs and hydrothermal vents that provide the high temperatures to which they are adapted. Organisms living deep below the surface of the Earth (see Chapter 2 – The underworld) also live at high temperatures and are widespread, with a total biomass perhaps exceeding that at the surface. Maybe these hyperthermophiles retreated to their subterranean world as the surface cooled, or perhaps life evolved there in the first place.

Is the deep sea extreme?

The deep sea is widespread. The seabed, covered by more than 3000 metres of seawater, makes up 53 per cent of the total surface of the Earth. The deep sea is likely to have been present for most or all of the time that organisms have been present on the Earth. For organisms that evolved in surface waters, life in the depths presented a number of problems. In particular, organisms which colonised the deep sea had to develop mechanisms for coping with the high pressures that are found there (see Chapter 6, Under pressure). A wide diversity of invertebrate animals inhabit the deep sea (see Chapter 2, The cold deep sea) and it is likely to have been colonised at different times by different groups of animals. Fishes are probably the most recent colonisers, with 10–15 per cent of fish species living there.

The deep sea clearly represented a challenge to its colonisers, but it also provided an opportunity. As a whole, it is the largest environment on the Earth's surface. It is supplied with food via the dead bodies, and other products, of organisms from surface waters. We might think of the colonisation of the deep sea as being comparable to the colonisation of the land. It was once an extreme environment, but, just as we would not now consider most terrestrial environments to be extreme, such a wide diversity of organisms have solved the problems of living in the deep sea that we cannot now consider it to be extreme. Once the problems of living in a widespread environment had been solved, colonising organisms would also have become widespread and developed into a diversity of species. They are encompassed by the life box of the majority of species. In other words, the life box of life on Earth has expanded with time. Some extreme environments, however, are rare and are likely to have been rare throughout the history of the Earth. Examples include salt lakes and extreme acidic or alkaline conditions. Organisms that colonise these environments will remain rare and their status as extreme organisms will not change. However, not all rare habitats are extreme. Extreme habitats are those whose physical conditions (temperature, salinity, water availability etc.) lie outside those experienced by the majority of organisms.

Pioneers and stragglers

We can envisage how extreme environments and extreme organisms have changed with time by considering how their life boxes, and the life box for life on Earth in general, have changed (Figure 8.7). We might recognise three situations: the first two involve extreme organisms being pioneers and the third involves them being stragglers. Extreme organisms that are pioneer colonists of a new (and extreme) but widespread environment will no longer be considered extreme once the problems of living in that environment have been solved and its colonisation has become widespread (Figure 8.7A). Pioneer colonists of rare extreme environments are likely to remain rare and extreme (Figure 8.7B). If the average conditions on Earth change, those organisms which do not adapt to the new conditions (the stragglers) may be able to survive in the now rare and extreme environments that retain the old conditions (Figure 8.7C). The life box can thus give us a way of recognising extreme organisms, but only if we realise that the situation is dynamic.

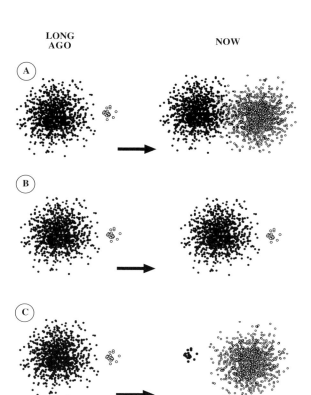

FIGURE 8.7   Three possible models for the evolution of extreme organisms. Open circles represent organisms that have changed to a new set of conditions, while closed circles represent those which have retained an old set of conditions. Those conditions and organisms which lie outside those of the majority are considered extreme.

In (A), the extreme organisms are pioneer organisms colonising an extreme but widespread environment (such as the land or the deep sea). As the problems of living in the new environment are conquered, organisms become widespread in this environment and can no longer be considered extreme, since their life boxes are encompassed by those of the majority of organisms.

In (B), the extreme organisms are pioneer organisms colonising an extreme and rare environment (such as salt lakes). The environment remains rare and the organisms continue to be considered extreme.

In (C), the conditions on Earth have changed (as in the change from anaerobic to aerobic conditions). Most species have evolved to survive in the new conditions and many have become extinct, but the stragglers survive in the now rare and extreme environments which retain the old conditions.

THE CHALLENGE OF THE EXTREME

The study of extreme organisms offers plenty of challenges to biologists as we struggle to understand the mechanisms by which they manage to survive, and even flourish, under conditions that would be fatal to most other organisms. As well as its intrinsic fascination, extreme biology has yielded plenty of practical outcomes and potential applications (see Chapters 3–6). Enzymes isolated from extreme organisms are used in processes ranging from laundry to DNA fingerprinting. Antifreezes from polar fish, and other cold-tolerant organisms, may improve our ability to cryopreserve biological materials and produce better frozen food, aid the storage of organs and tissues for transplantation, and yield smoother, creamier ice cream. Understanding the mechanisms of anhydrobiosis may also improve our ability to store biological materials. Crop plants may be grown in a wider range of environments if their ability to survive desiccation and temperature stress can be improved.

By understanding how life survives at the extremes, we may gain more understanding of how organisms in general survive and function in their environments, how life evolves and how organisms interact to form functioning ecological communities. We may also improve our understanding of the nature and origin of life and develop techniques for our search for life elsewhere in the universe.

Extreme biologists have tended to focus their attention on extremophiles (especially archaea and bacteria) that use capacity adaptation, in the form of enzymes, membranes and other components that withstand the extreme conditions. Cryptobiotic organisms, which provide an extreme example of resistance adaptation, have received much less attention and yet this phenomenon challenges our understanding of the nature of life itself. New discoveries of extreme organisms and their survival abilities and mechanisms are being reported almost daily. However, much remains that challenges our understanding and no doubt many surprises still await us.

Bibliography

General

I used the following as general sources of information and inspiration

Campbell, N.A. (1996). *Biology*, 4th edition. Menlo Park, CA: Benjamin/Cummings.

Encyclopaedia Britannica® DVD 2000 ©1994–2000. Encyclopaedia Britannica, Inc.

Madigan, M.T., Martinko, J.M. & Parker, J. (2000). *Brock Biology of Microorganisms,* 9th edition. Upper Saddle River, NJ: Prentice Hall.

Postgate, J. (1994). *The Outer Reaches of Life.* Cambridge: Cambridge University Press.

Wilmer, P., Stone, G. & Johnston, I. (2000). *Environmental Physiology of Animals.* Oxford: Blackwell Science.

The internet

Some websites to try include:

NASA Astrobiology Institute Website
  http://nai.arc.nasa.gov/

NASA's own Astrobiology website has information on organisms and extremes
  http://astrobiology.arc.nasa.gov/

Details of the Life in Extreme Environments Programme (US National Science Foundation) are at
  http://www.nsf.gov/home/crssprgm/lexen/start.htm

Astrobiology Web, Life in Extreme Environments (with numerous links)
  http://www.astrobiology.com/extreme.html

Explore the deep sea, courtesy of the University of Delaware
  http://www.ocean.udel.edu/deepsea/

Try this site for the deserts of the USA
  http://www.desertusa.com/index.html

The University of Canterbury's Gateway Antarctica site has many links to Antarctic resources on the web
  http://icair.iac.org.nz/GAhome.htm

The Roald Amundsen Centre for Arctic Research has links to Arctic sites
  http://www.arctic.uit.no/English/

1 – Introduction: extreme life

Brown, D.H. (1956). *Composition of Scientific Words*. Washington: Smithsonian Institution Press.

Hutchinson, G.E. (1957). Concluding remarks. *Cold Spring Harbor Symposium on Quantitative Biology* **22**, 415–27. [The ecological niche.]

Kennedy, M.J., Reader, S.L. & Swierczynki, L.M. (1994). Preservation records of micro-organisms: evidence of the tenacity of life. *Microbiology* **140**, 2513–29.

Parkes, R.J. (2000). A case of bacterial immortality? *Nature* **407**, 844–5.

Precht, H. (1958). Concepts of the temperature adaptation of unchanging reaction systems of cold-blooded animals. In *Physiological Adaptation*, ed. C.L. Prosser, pp. 351–76. Washington: American Association for the Advancement of Science. [Capacity and resistance adaptation.]

2 – Be it ever so humble...

Deserts, temporary waters and saline and soda lakes

Burgis, M.J. & Morris, P. (1987). *The Natural History of Lakes*. Cambridge: Cambridge University Press.

Crawford, C.S. (1981). *Biology of Desert Invertebrates*. Berlin: Springer-Verlag.

Gauthier-Pilters, H. & Dagg, A.I. (1981). *The Camel. Its Evolution, Ecology, Behaviour and Relationship to Man.* Chicago and London: University of Chicago Press.

Louw, G.N. & Seeley, M.K. (1982). *Ecology of Desert Organisms*. London and New York: Longman

Sømme, L. (1995). *Invertebrates in Hot and Cold Arid Environments*. Berlin: Springer-Verlag.

Wagner, F.H. (1980). *Wildlife of the Deserts*. New York: Abrahams Inc.

Watzman, H. (1997). Left for dead. *New Scientist*, 8 February 1997, pp. 37–41. [The Dead Sea.]

Williams, D.D. (1987). *The Ecology of Temporary Waters*. London and Sydney: Croom Helm.

Overwintering, polar and alpine environments

Bonner, W.N. & Walton, D.W.H. (1985). *Key Environments – Antarctica*. Oxford and New York: Pergamon Press,.

Chernov, Y.I. (1985). *The Living Tundra*. Cambridge: Cambridge University Press.

Fogg, G.E. (1998). *The Biology of Polar Habitats.* Oxford: Oxford University Press.

Laws, R.M. (1984). *Antarctic Ecology*. London: Academic Press.

Leather, S.R., Walters, K.F.A. & Bale, J.S. (1993). *The Ecology of Insect Overwintering*. Cambridge: Cambridge University Press.

Lyman, C.P., Willis, J.P., Malan, E. & Wang, L.C.H. (1982). *Hibernation and Torpor in Mammals and Birds.* London: Academic Press.

Mani, M.S. (1980). *Ecology of Highlands.* The Hague: W Junk bv Publishers.

Sage, B.L. (1986). *The Arctic and its Wildlife.* New York and Oxford: Facts on File Publications.

Stonehouse, B. (1971). *Animals of the Arctic: the Ecology of the Far North.* London: Ward Lock.

Stonehouse, B. (1972). *Animals of the Antarctic: The Ecology of the Far South.* London: Peter Lowe.

Vincent, W.F. (1988). *Microbial Ecosystems of Antarctica.* Cambridge: Cambridge University Press.

In the depths

Fredrickson, J.K. & Onstott, T.C. (1996). Microbes deep inside the earth. *Scientific American,* October 1996, pp. 42–7.

Pain, S. (ed.) (1996). Unknown oceans. *New Scientist supplement,* 2 November 1996.

Pain, S. (1998). The intraterrestrials. *New Scientist,* 7 March 1998, pp. 28–32.

Robinson, B.H. (1995). Light in the ocean's midwaters. *Scientific American,* July 1995, pp. 50–6.

Suess, E., Bohrmann, G., Greinert, J. & Lausch, E. (1999). Flammable ice. *Scientific American,* November 1999, pp. 52–9. [Methane hydrates.]

Van Dover, C.L. (2000). *The Ecology of Deep-Sea Hydrothermal Vents.,* Princeton, NJ: Princeton University Press.

Life in life

Douglas, A.E. (1994). *Symbiotic Interactions.* Oxford: Oxford University Press.

Matthews, B.E. (1998). *An Introduction to Parasitology.* Cambridge: Cambridge University Press.

3 – Life without water

Barrett, J. (1982). Metabolic responses to anabiosis in the fourth stage juveniles of *Ditylenchus dipsaci* (Nematoda). *Proceedings of the Royal Society of London* **B216**, 159–77.

Barrett, J. (1991). Anhydrobiotic nematodes. *Agricultural Zoology Reviews* **4**, 161–76.

Clegg, J.S. (1979). Metabolism and the intracellular environment: the vicinal-water network model. In *Cell Associated Water,* eds. W. Drost-Hansen & J.S. Clegg, pp. 363–413. New York: Academic Press. [The state of water and metabolism in desiccating *Artemia* cysts.]

Close, T.J. (1996). Dehydrins: emergence of a biochemical role of a family of plant dehydration proteins. *Physiologia Plantarum* **97**, 795–803.

Copley, J. (1999). Indestructible. *New Scientist*, 23 October 1999, pp. 44–6. [Tardigrades.]

Crowe, J.H. (1971). Anhydrobiosis: an unsolved problem. *American Naturalist* **105**, 563–73.

Crowe, J.H. & Cooper, A.E. (1971). Cryptobiosis. *Scientific American*, December 1971, pp. 30–6.

Crowe, J.H., Carpenter, J.F. & Crowe, L.M. (1998). The role of vitrification in anhydrobiosis. *Annual Review of Physiology* **60**, 73–103.

Crowe, J.H., Hoekstra, F. & Crowe, L.M. (1992). Anhydrobiosis. *Annual Review of Physiology* **54**, 579–99.

Gaff, D.F. (1977). Desiccation tolerant vascular plants of Southern Africa. *Oecologia* **31**, 95–109.

Gaff, D.F. (1997). Mechanisms of desiccation tolerance in resurrection vascular plants. In *Mechanisms of Environmental Stress Resistance in Plants*, eds. A.S. Basra & R.K. Basra, pp. 43–58. Amsterdam: Harwood Academic Publishers.

Hinton, H.E. (1960). Cryptobiosis in the larva of *Polypedilium vanderplanki* Hint. *Journal of Insect Physiology* **5**, 286–300.

Holmström, K., Mantyla, E., Welin, B., Mandal, A. & Palva E.T. (1996). Drought tolerance in tobacco. *Nature* **379**, 683–4. [Genetic modification by the addition of genes for trehalose synthase.]

Horneck, G. (1998). Exobiological experiments in Earth orbit. *Advances in Space Research* **22**, 317–26. [Bacterial spores etc. in space.]

Ingram, J. & Bartels, D. (1996). The molecular basis of dehydration tolerance in plants. *Annual Review of Plant Physiology and Plant Molecular Biology* **47**, 377–403.

Keilin, D. (1959). The Leeuwenhoek Lecture. The problem of anabiosis or latent life: history and current concept. *Proceedings of the Royal Society of London* **B150**, 149–91.

Oliver, M.J., Wood, A.J. & O'Mahony, P. (1998). 'To dryness and beyond' – preparation for the dried state and rehydration in vegetative desiccation-tolerant plants. *Plant Growth Regulation* **24**, 193–201.

Potts, M. (1994). Desiccation tolerance of prokaryotes. *Microbiological Reviews* **58**, 755–805.

Roser, B. & Colaço, C. (1993). A sweeter way to fresher food. *New Scientist*, 15 May 1993, pp. 25–8.

Virginia, R.A. & Wall, D.H. (1999). How soils structure communities in the Antarctic Dry Valleys. *BioScience* **49**, 973–83. [Nematodes in the soil of the Dry Valleys.]

Wharton, D.A. (2002). Survival strategies. In *The Biology of Nematodes*, ed. D.L. Lee pp. 389–411. London: Taylor and Francis.

Wharton, D.A., Barrett, J. & Perry, R.N. (1985). Water uptake and morphological changes during recovery from anabiosis in the plant-parasitic nematode, *Ditylenchus dipsaci. Journal of Zoology (London)* **206**, 391–402.

Womersley, C. (1987). A reevaluation of strategies employed by nematode anhydrobiotes in relation to their natural environment. In *Vistas on Nematology*, eds. J.A. Veech & D.W. Dickson, pp. 165–73. Hyattsville, MD: Society of Nematologists Inc.

Womersley, C.Z., Wharton, D.A. & Higa, L.M. (1998). Survival biology. In *The Physiology and Biochemistry of Free-Living and Plant-Parasitic Nematodes*, Eds. R.N. Perry & D.J. Wright, pp. 271–302. Wallingford and New York: CABI Publishing.

Wright, J.C., Westh, P. & Ramløv, H. (1992). Cryptobiosis in Tardigrada. *Biological Reviews* **67**, 1–29.

Young, S.R. (1985). The dry life. *New Scientist*, 31 October 1985, pp. 40–4.

4 – The hot club

Brock, T.D. (1978). *Thermophilic Microorganisms and Life at High Temperatures.* Berlin: Springer-Verlag.

Cary, S.C., Shank, T. & Stein, J. (1998). Worms bask in extreme temperatures. *Nature* **391**, 545–6. [Pompeii worms.]

Cossins, A.R. & Bowler, K. (1987). *Temperature Biology of Animals.* London and New York: Chapman and Hall.

Desbruyères, D., Chevaldonné, P., Alayse, A.-M. *et al.* (1998). Biology and ecology of the 'Pompeii worm' (*Alvinella pompejana* Desbruyères and Laubier), a normal dweller of an extreme deep-sea environment: a synthesis of current knowledge and recent developments. *Deep-Sea Research II* **45**, 383–422.

Morimoto, R.I., Tissières, A. & Georgopoulos, C. (1994). *The Biology of Heat Shock Proteins and Molecular Chaperones.* Cold Spring Harbor, NY: Cold Spring Harbor Laboratory Press.

Schmidt-Nielsen, K., Taylor, C.R. & Shkolnik, A. (1971). Desert snails: problems of heat, water and food. *Journal of Experimental Biology* **55**, 385–98.

Seckbach, J. (ed.) (1999). *Enigmatic Microorganisms and Life in Extreme Environments.* Dordrecht: Kluwer Academic Publishers

Wilmer, P., Stone, G. & Johnston, I. (2000). *Environmental Physiology of Animals*, Chapter 8. Oxford: Blackwell Science.

5 – Cold Lazarus

Block, W. (1995). Insects and freezing. *Science Progress* **78**, 349–72.

Cannon, R.J.C. & Block, W. (1988). Cold tolerance of microarthropods. *Biological Reviews* **63**, 23–77. [Includes Antarctic mites and springtails.]

Carpenter, H. (1998). *Dennis Potter: a Biography*. London: Faber and Faber.

Costanzo, J.P. & Lee, R.E. (1994). Biophysical and physiological responses promoting freeze tolerance in vertebrates. *News in Physiological Sciences* **9**, 252–6.

Davenport, J. (1992). *Animal Life at Low Temperatures*. London and New York: Chapman and Hall.

Goodman, B. (1998). Where ice isn't nice: how the icefish got its antifreeze and other tales of molecular evolution. *Bioscience* **48**, 586–90.

Holmstrup, M. & Zachariassen, K.E. (1996). Physiology of cold hardiness in earthworms. *Comparative Biochemistry and Physiology A* **115**, 91–101.

Hoshino, T., Odaira, M., Yoshida, M. & Tsuda, S. (1999). Physiological and biochemical significance of antifreeze substances in plants. *Journal of Plant Research* **112**, 255–61.

Knight, J. (1998). Life on ice. *New Scientist*, 2 May 1998, pp. 24–8.

Lee, R.E. & Costanzo, J.P. (1998). Biological ice nucleation and ice distribution in cold-hardy ectothermic animals. *Annual Review of Physiology* **60**, 55–72.

Lee, R.E. & Denlinger, D.L. (1991). *Insects at Low Temperatures*. London and New York: Chapman and Hall.

Li, P.H. & Chen, T.H.H. (1996). *Plant Cold Hardiness: Molecular Biology, Biochemistry and Physiology*. New York and London: Plenum Press.

Margesin, R. & Schinner, F. (1999). *Cold-Adapted Organisms: Ecology, Physiology, Enzymology and Molecular Biology*. Berlin: Springer-Verlag.

Margesin, R. & Schinner, F. (1999). *Biotechnological Applications of Cold-Adapted Organisms*. Berlin: Springer-Verlag.

Ramløv, H. (2000). Aspects of natural cold tolerance in ectothermic animals. *Human Reproduction* **15**, 26–46.

Sakai, A. & Larcher, W. (1987). *Frost Survival of Plants*. Berlin: Springer-Verlag.

Smith, A.U. (1961). *Biological Effects of Freezing and Supercooling*. London: Edward Arnold. [For historical accounts.]

Sømme, L. (1995). *Invertebrates in Hot and Cold Arid Environments*. Berlin: Springer-Verlag.

Sømme, L. (2000). The history of cold hardiness research in terrestrial arthropods. *Cryo Letters* **21**, 289–96.

Storey, K.B. (1999). Living in the cold: freeze-induced gene responses in freeze-tolerant vertebrates. *Clinical and Experimental Pharmacology and Physiology* **26**, 57–63.

Storey, K.B. & Storey, J.M. (1990). Frozen and alive. *Scientific American*, December 1990, pp. 62–7.

Storey, K.B. & Storey, J.M. (1992). Natural freeze tolerance in ectothermic vertebrates. *Annual Review of Physiology* **54**, 619–37.

Vincent, W.F. (1988). *Microbial Ecosystems of Antarctica*. Cambridge: Cambridge University Press.

Wharton, D.A. (2002). Survival strategies. In *The Biology of Nematodes* ed. D.L. Lee, pp. 389–411. London: Taylor and Francis.

Wharton, D.A. & Ferns, D.J. (1995). Survival of intracellular freezing by the Antarctic nematode *Panagrolaimus davidi*. *Journal of Experimental Biology* **198**, 1381–7.

6 – More tough choices

Abe, F., Kato, C. & Horikoshi, K. (1999). Pressure-regulated metabolism in microorganisms. *Trends in Microbiology* **7**, 477–53.

Bartlett, D.H. (1992). Microbial life at high pressures. *Science Progress* **76**, 479–96.

Horikoshi, K. & Grant, W.D. (Eds.) (1998). *Extremophiles: Microbial Life in Extreme Environments*. New York: Wiley-Liss.

Horikoshi, K. & Tsujii, K. (Eds.) (1999). *Extremophiles in Deep-Sea Environments*. Berlin: Springer-Verlag.

Kushner, D.J. (Ed.) (1978). *Microbial Life in Extreme Environments*. London: Academic Press.

Madigan, M.T., Martinko, J.M. & Parker, J. (2000). *Brock Biology of Microorganisms*, 9th edition. Upper Saddle River, NJ: Prentice Hall.

Postgate, J. (1994). *The Outer Reaches of Life*. Cambridge: Cambridge University Press.

Randall, D.J. & Farrell, A.P. (eds.) (1997). *Deep-Sea Fishes*. Academic Press, London.

Schleper, C., Pühler, G., Kühlmorgan, B. & Zillig, W. (1995). Life at extremely low pH. *Nature* **375**, 741–2.

Seckbach, J. (Ed.) (1999). *Enigmatic Microorganisms and Life in Extreme Environments*. Dordrecht: Kluwer Academic Publishers.

7 – 'It's life, Jim, but not as we know it!'

Boyce, N. (1999). The Martians in your kidneys. *New Scientist*, 21 August 1999, pp. 32–35. [Nanobes and nanobacteria.]

Brack, A. (Ed.) (1998). *The Molecular Origins of Life: Assembling Pieces of the Puzzle*. Cambridge: Cambridge University Press.

Davies, P.C.W. (1999). Life force. *New Scientist*, 18 September 1999, pp. 27–30. [Origin of life.]

Davies, P.C.W. (1999). *The Fifth Miracle: The Search for the Origin and Meaning of Life.* New York: Simon and Schuster.

Falk, D. (1999). Alien haven. *New Scientist*, 18 September 1999, pp. 32–5. [extraterrestrial life.]

Hinton, H.E. & Blum, M.S. (1963). Suspended animation and the origin of life. *New Scientist*, 28 October 1965, pp. 270–1.

Jakosky, B. (1998). *The Search for Life on Other Planets.* Cambridge: Cambridge University Press.

Malin, M.C. & Edgett, K.S. (2000). Evidence for recent groundwater seepage and surface runoff on Mars. *Science* **288**, 2330–5.

Maynard Smith, J. & Szathmary, E. (1999). *The Origins of Life : From the Birth of Life to the Origin of Language.* Oxford: Oxford University Press.

Tater, J.C. & Chyba, C.F. (1999). Is there life elsewhere in the Universe? *Scientific American*, December 1999, pp. 80–5.

Walker, G. (1999). Waterworld. *New Scientist*, 18 September 1999, pp. 41–3. [Europa.]

Water, M. (1999). *The Search for Life on Mars.* Cambridge, MA: Perseus Books.

8 – An extreme biology

Bayley, M. & Holmstrup, M. (1999). Water vapour absorption in arthropods by the accumulation of myoinositol and glucose. *Science* **285**, 1909–11.

Bayly, I.A.E. & Williams, W.D. (1973). *Inland Waters and Their Ecology.* Hawthorn, VIC: Longman Australia.

Begon, M., Harper, J.L. & Townsend, C.R. (1996). *Ecology: Individuals, Populations and Communities.* Oxford: Blackwell Science.

Cheng, C.-H.C. (1998). Evolution of the diverse antifreeze proteins. *Current Opinion in Genetics and Development* **8**, 715–20.

Cheng, C.-H.C. & Chen, L. (1999). Evolution of an antifreeze glycoprotein. *Nature* **401**, 443–4.

Clarke, A. & Jonson, I.A. (1996). Evolution and adaptive radiation of Antarctic fishes. *Trends in Ecology and Evolution* **11**, 212–18.

Convey, P. (1996). The influence of environmental characteristics on life history attributes of Antarctic terrestrial biota. *Biological Reviews* **71**, 191–225.

Feder, M.E. (1999). Organismal, ecological, and evolutionary aspects of heat-shock proteins and the stress response: established conclusions and unresolved issues. *American Zoologist* **39**, 857–64.

Goodman, B. (1998). Where ice isn't nice: how the icefish got its antifreeze and other tales of molecular evolution. *BioScience* **48**, 586–90.

Hoffmann A.A. & Hercus, M.J. (2000). Environmental stress as an evolutionary force. *BioScience* **50**, 217–26.

Kempf, B. & Bremer, E. (1998). Uptake and synthesis of compatible solutes as microbial stress responses to high-osmolality environments. *Archives of Microbiology* **170**, 319–30.

Little, C. (1983). *The Colonisation of Land: Origins and Adaptations of Terrestrial Animals*. Cambridge: Cambridge University Press

Logsdon, J.A. & Doolittle, W.F. (1997). Origin of antifreeze protein genes: a cool tale in molecular evolution. *Proceedings of the National Academy of Sciences* **94**, 3485–7.

Pain, S. (1996). Monster journeys. *New Scientist supplement*, 2 November 1996, pp. 15–17. [dispersal of hydrothermal vent fauna.]

Tunnicliffe, V., McArthur, A.G. & McHugh, D. (1998). A biogeographical perspective of the deep-sea hydrothermal vent fauna. *Advances in Marine Biology* **34**, 353–442.

Waterman, T.H. (1999). The evolutionary challenges of extreme environments (part 1). *Journal of Experimental Zoology* **285**, 326–59.

Glossary

abscisic acid–a plant hormone

absolute zero–the lowest temperature theoretically obtainable (−273 °C)

Acanthocephala–a phylum of parasitic invertebrate animals

acclimation–a physiological adjustment to a change in a physical factor induced in the laboratory

acclimatisation–a physiological adjustment to a change in an environmental factor (in nature)

acidophile–an organism that grows best at low pH

acidotolerant–an organism that can tolerate low pH but which grows best at a higher pH

actinomycetes–a group of bacteria noted for their filamentous and branching growth patterns

adaptation–an evolutionary change by which an organism becomes better suited to its environment

aerobic–an environment that contains, or an organism or process that requires, oxygen

alga (plural: algae)–a photosynthetic, plant-like protist

alkaliphile–an organism that grows best at high pH

alkalitolerant–an organism that can tolerate high pH but which grows best at a lower pH

ametabolic–no metabolism

amino acids–organic compounds that form the building blocks of proteins

amphiphilic–a molecule that has both hydrophilic and hydrophobic parts

anabiosis–see cryptobiosis

anaerobic–an environment that lacks oxygen, or an organism or process that requires the absence of oxygen

angiosperm – a flowering plant

anhydrobiosis – surviving a cessation of metabolism due to water loss

animalcules – microscopic animals (archaic term)

annelids – a phylum of segmented worm-like invertebrate animals (includes earthworms)

anoxic – lacking oxygen

anoxybiosis – surviving a cessation of metabolism due to the absence of oxygen

Antarctic Circumpolar Current – an ocean current that completely encircles Antarctica, isolating it from the oceans of the rest of the world

Antarctic Polar Front – the region of the Antarctic Ocean where the cold, nutrient-rich surface waters of the Antarctic sink beneath the warmer northern waters

antifreeze – a substance that lowers the freezing point of an organism's body fluids

antifreeze proteins – proteins that inhibit the growth of ice crystals by attaching to their surface

arachnids – a group of arthropods that includes scorpions, spiders, mites, ticks, harvest-men and king crabs

Archaea – microorganisms that are similar in size and appearance to bacteria but very different in their molecular organisation. They form the third domain of life (the others being Bacteria and Eukarya)

arthropods – a phylum of invertebrate animals with a hard exoskeleton and jointed legs (includes insects, arachnids, crustaceans and millipedes)

assimilation – the incorporation of food and/or inorganic materials into the complex constituents of an organism

astrobiology (or exobiology) – the branch of biology that deals with the possibility, and likely nature, of extraterrestrial life

atmosphere (unit of measurement) – the average atmospheric pressure at sea level

autotroph – an organism that is able to produce organic compounds from simple inorganic compounds

bacterium (plural: bacteria) – a unicellular, prokaryotic microorganism that lacks an organised nucleus and organelles

barophilic – see piezophilic

barotolerant – see piezotolerant

benthic – living at the bottom of a sea or lake

biomass – the total mass of organisms in a given area

bound water – see osmotically inactive water

bryophyte (moss) – a small flowerless green plant that lacks true roots

capacity adaptation – the ability to grow and reproduce under extreme environmental conditions

carbohydrates – a group of organic compounds that occurs in living organisms (includes sugars, starch and cellulose)

catalyst – a substance that increases the rate of a chemical reaction without itself undergoing any permanent change

chaperones – see molecular chaperones

chemotroph – an organism that is able to produce organic compounds from simple inorganic compounds by obtaining energy from the oxidation of inorganic substances

chilling intolerance – the death of an organism from the effects of cold but before it freezes (prefreeze mortality)

chironomids (midges) – a group of insects

chlorophyll – the green pigment of plants and cyanobacteria that absorbs light and which provides the energy for photosynthesis

chloroplasts – the organelles of algae and plant cells in which the processes of photosynthesis occur

chorion – eggshell (particularly of fish and insect eggs)

cloaca – the common opening of the digestive, urinary and reproductive systems in some animals

cold coma – the activity of an organism ceasing at low temperatures

cold hardening – the process by which an organism increases its cold tolerance

cold shock – inactivity induced by cold

cold tolerance – the ability to survive low temperatures

collembola – see springtails

colligative properties–properties that depend only on the concentration of a solute and not on its composition

compatible solutes–a solute that does not adversely affect the workings of cells

concentration–the relative amount of a substance contained within a particular volume of space, e.g. the amount of solute per unit volume of solution

conformation–the three-dimensional shape of a molecule

consumers–organisms that satisfy their requirements for organic molecules and energy by eating (or otherwise utilising the products of) autotrophs (or other consumers)

convergent evolution–the independent evolution of similar characteristics in two groups of organisms as a result of similar selection pressures

crustacean–a group of arthropod animals that includes shrimps, crabs, water fleas etc.

cryobiosis–surviving a cessation of metabolism due to low temperatures

cryogenics–the production and application of low temperatures. Also refers to the freezing of human bodies in the hope of being able to revive them at some future date

cryopreservation–the preservation of biological materials (such as food or human tissues) by freezing

cryoprotectant–a substance that protects against the harmful effects of freezing

cryptobiosis–surviving a cessation of metabolism

cuticle–a protective layer covering the outside of a plant or invertebrate animal

cyanobacteria–a group of photosynthetic bacteria

cyst–a tough protective capsule which contains the resting stage of an organism

cytoplasm–the contents of a cell, excluding the nucleus

deciduous–a plant that sheds its leaves annually

decomposers–organisms that get their food from dead organisms or their waste products

dehydrins – proteins in plants that are induced by desiccation

denaturation – a process by which a protein unravels and loses its natural conformation, destroying its biological activity

detritus – dead organisms and other organic debris

diapause – a period of suspended development

diatoms – single-celled algae with a cell wall of silica

diffusion – the spontaneous movement of a substance from a region of high concentration to one of low concentration

DNA (deoxyribonucleic acid) – a self-replicating organic molecule found in nearly all organisms which is the main constituent of chromosomes and whose sequence of nucleotides carries the organism's genetic information

domain – the highest taxonomic grouping of organisms. There are three domains: Bacteria, Archaea and Eukarya

dormancy – a period during which the normal physiological processes of an organism are slowed down or suspended

ecological niche – the sum total of the physical and biological characteristics that determine the place an organism occupies in an ecological community

ectotherm – an organism that is dependent on external sources of heat (usually the sun)

enchytraeids – a group of annelid worms

endolithic – living within rocks

endoplasmic reticulum – an extensive network of membranes within the cytoplasm of eukaryotic cells

endotherm – an organism that is able to generate heat internally

entropy – a measure of the degree of randomness or disorder in a system

enzyme – a protein that catalyses (increases the rate of) biological reactions

Eukarya – the domain of eukaryotic organisms (fungi, protists, plants and animals)

eukaryote – an organism which has a cell, or cells, where the genetic material is contained within a distinct membrane-bound nucleus and which has membrane-bound organelles

evolution (Darwinian evolution)–the process by which organisms develop and diversify from organisms that precede them

exobiology–see astrobiology

extracellular–outside the cells of an organism

extradural–outside the membrane that encloses the brain and spinal cord

extreme biology–the study of extreme organisms and extreme environments

extremophile–an organism that will grow and reproduce under extreme environmental conditions

extremozymes–enzymes that operate under extreme conditions

fatty acids–one of the subunits of lipids

fern–a plant that lacks flowers and which reproduces by producing spores

freeze avoiding–an organism that survives low temperature by preventing ice from forming within its body

freeze concentration–the concentration of salts in the remaining unfrozen portion of a solution as a result of the sequestration of water molecules into ice

freezing tolerant–an organism that survives ice forming within its body

fungi (singular: fungus)–multicellular or unicellular eukaryotes that absorb their food across their cell walls after secreting enzymes onto it

glycogen–the main carbohydrate store in animals

glycoprotein–a protein with an attached carbohydrate group

gymnosperms–plants that produce unprotected seeds (e.g. conifers)

habitable zone–the zone around a star where planets capable of supporting life may be found

haemoglobin–the red oxygen-transporting pigment of the blood

haemolymph–the body fluid of arthropods (equivalent to the blood of vertebrates)

halobacteria–salt-tolerant archaea

halophile–an organism that grows best at high salt concentrations

halotolerant–an organism that tolerates high salt concentrations but grows best at lower concentrations

heat coma–the activity of an organism ceasing at high temperatures

heat of crystallisation–see latent heat of crystallisation

heat shock response–the induction of protective proteins, such as molecular chaperones, by a mild heat shock

heterotroph–an organism that requires complex organic molecules (usually obtained by feeding on other organisms)

hibernation–spending the winter in a dormant state

histones–a group of small proteins that are associated with DNA in chromosomes

homeostasis–the maintenance of a constant environment within an organism by physiological mechanisms

hydrophilic–water attracting

hydrophobic–water repelling

hydrothermal vent–an opening in the sea floor through which geothermally heated water flows

hyperosmotic stress–an osmotic stress produced when a cell or organism is immersed in a solution which contains a higher concentration of solutes than that within its own cells or fluids

hyperthermophiles–organisms that grow at very high temperatures (between 80 °C and 100 °C)

hyposmotic stress–an osmotic stress produced when a cell or organism is immersed in a solution which contains a lower concentration of solutes than that within its own cells or fluids

ice-active protein–a protein that affects the formation and/or stability of ice

ice-nucleating protein–a protein that acts as an ice nucleator

ice nucleators–substances that cause ice nucleation (see nucleation)

impermeable–a membrane (or other structure) is impermeable to a substance if it does not allow that substance to pass through it

inoculative freezing–the freezing of an organism as a result of ice from its surroundings travelling across its surface

inorganic compounds–originally referred to chemical compounds that did not derive from living organisms but now generally refers to chemical compounds that do not contain carbon

intracellular–inside the cells of an organism

invertebrate–an animal without a backbone

ion–an atom or molecule with a net electrical charge

kinetic energy–a form of energy possessed by an object by virtue of its motion

kingdom–the highest taxonomic classification below the domain (e.g. plants, animals)

late embryogenesis-abundant proteins–proteins that are produced by plant seeds during the later stages of their development

latent heat of crystallisation (latent heat of fusion)–the heat released during the conversion of a substance from a liquid to a solid (as in the freezing of water)

lichen–a symbiotic association between a fungus and an alga or cyanobacterium, which forms an encrusting or plant-like growth

lipid–a group of organic compounds that are insoluble in water (includes fats, phospholipids, oils and waxes)

liverworts–small flowerless plants that lack roots and which reproduce by producing spores

lower lethal temperature–the lowest temperature that an organism will survive

membrane–a sheet-like layer that forms partitions within a cell, and between a cell and its surroundings, and which consists of a double layer of lipids with associated proteins and carbohydrates

metabolism–the chemical processes that occur within an organism and which keep it alive

methanogen–an organism that produces methane

microorganism–refers to any organism that is too small to be seen with the naked eye; usually used in relation to bacteria, archaea, fungi, algae, viruses and protists

mitochondrion (plural: mitochondria)–the organelle of eukaryotic cells in which the processes of respiration and energy production occur

molecular chaperones–proteins that assist in the correct folding of other proteins

molluscs–a phylum of invertebrate animals, including snails, slugs, octopus and squid

Monera–a term sometimes used for all prokaryotic organisms

moss–see bryophyte

multicellular–consisting of many cells

mutation–a change in the structure of a gene

mutualism–a symbiotic relationship in which both partners in the association benefit

nematodes–a phylum of worm-like invertebrate animals that are parasitic in animal or plants, or free-living in soil and in freshwater and marine sediments

niche–see ecological niche

nucleation (of ice formation)–the initial process which results in the formation of an ice crystal

nucleic acids (see DNA and RNA)–organic molecules involved in the reproduction of organisms and which consist of subunits called nucleotides

nucleotide–one of the basic structural units of nucleic acids (DNA and RNA)

nucleus–a membrane-bound organelle that contains the genetic material of a eukaryotic cell

organelle–refers to a variety of organised and specialised structures (often membrane bound) found in cells

organic compounds–originally referred to chemical compounds that derived from living organisms but now generally refers to chemical compounds that contain carbon

osmobiosis–surviving a cessation of metabolism due to osmotic stress

osmolyte–an osmotically active substance

osmosis–the process by which molecules of a solvent tend to pass through a semi-permeable membrane from a less concentrated solution into a more concentrated one

osmotic stress–see hyperosmotic stress and hyposmotic stress

osmotically inactive water–water that is not free to move under an osmotic stress (also sometimes called 'bound water' or 'unfreezable water')

panspermia–the theory that life on Earth arose from microorganisms travelling through space

parasitism–an association between two organisms in which one partner (the parasite) causes harm to the other (the host)

pelagic–living in the open sea

permafrost–permanently frozen ground

permeable–a membrane (or other structure) is permeable to a substance if it allows that substance to pass through it

pH–a measure of the acidity or alkalinity of a solution (the concentration of hydrogen ions on a logarithmic scale)

phospholipid–a lipid containing a phosphate group as part of its molecule

photoperiod–the relative length of night and day

photosynthesis–the process by which plants, and some other organisms, use the energy from sunlight to produce sugars from carbon dioxide and water

phototroph–an organism that is able to produce organic compounds from simple inorganic compounds by using the energy from sunlight

phylum (plural: phyla)–a taxonomic category used in the classification of organisms (ranks above class and below kingdom)

phytoplankton–microscopic plants and algae drifting in sea or freshwater

piezophile–an organism that grows best at high pressures

piezotolerant–an organism that tolerates high pressure but which grows best at lower pressures

plasma membrane–the outer membrane which encloses a cell

pogonophorans–a phylum of worm-like marine invertebrate animals

polychaetes–a group of annelid worms

polyol–a sugar alcohol (e.g. glycerol, sorbitol)

polysaccharides–large molecules consisting of repeating sugar subunits

prebiotic – before the emergence of life

prefreeze mortality – see chilling intolerance

primary producer – organisms that use the energy from sunlight or chemical processes to produce organic material from inorganic compounds

productivity – the rate of production of new biomass by an organism, population or ecological community

prokaryote – an organism whose cell lacks a membrane-bound nucleus and organelles

proteins – large molecules consisting of chains of amino acids, which are essential components of the structure and function of all organisms

protists – eukaryotic organisms that are not included among the animals, plants or fungi; most are unicellular

protoplanetary disc – a flattened disc of dust around a star that may eventually coalesce to form planets

protozoa – animal-like protists

psychrophile – an organism that grows best at low temperatures

psychrotolerant – tolerating low temperatures but growing best at higher temperatures

quiescence – a period of dormancy

radical – a molecule that contains at least one unpaired electron

recrystallisation (of ice) – a change in the size and/or shape of ice crystals

recrystallisation inhibitor – a substance that inhibits recrystallisation

rehydrins – proteins in plants that are induced by rehydration

relative humidity – a measure of the degree of dryness or wetness of air (the amount of water vapour present in air expressed as a percentage of the amount needed to saturate it with water at the same temperature)

remote-operated vehicle (ROV) – an unmanned submarine operated by cable from a ship on the surface of the sea

renaturation – the restoration of the shape or conformation of a protein and the recovery of its biological function

resistance adaptation–the survival of extreme environmental conditions by a period of dormancy

resurrection plants–desiccation-tolerant higher plants

RNA (ribonucleic acid)–a nucleic acid found in all cells whose main role is to carry the instructions from DNA for the synthesis of proteins

rotifers (wheel animalcules)–a phylum of microscopic invertebrate animals; the name refers to a circle of cilia around the head, the beating of which gives the appearance of a rotating wheel

salinity–the concentration of salts in a solution

saltern–a pool where seawater is left to evaporate to make salt

saturated fatty acids–fatty acids that do not contain double bonds

semi-permeable membrane–a membrane with pores that allows the passage of small molecules (such as water) but not large molecules

slime mould–a simple organism (protist) that consists of a creeping mass of jelly-like protoplasm or cells

solute–a substance that is dissolved in a solution

solution–a liquid mixture of two or more substances

species richness–the number of species present in a given area

spiracle–the respiratory opening of an insect or other arthropod

spore–the resistant stage of a microorganism or plant

springtails (collembolans)–a group of small wingless insects

stomata–pores in the leaves and stems of plants that allow gas exchange

supercooling–the maintenance of a fluid in a liquid state at temperatures below its melting point

supercooling point–the temperature at which a supercooled fluid freezes

symbiosis–a close association between two organisms

tardigrades (water bears)–a phylum of microscopic invertebrate animals

temperature of crystallisation–see supercooling point

thermal hysteresis–a difference between the melting point and the freezing point of a solution, in the presence of an ice crystal

thermal hysteresis proteins–see antifreeze proteins

thermobiosis–surviving a cessation of metabolism due to high temperatures

thermophile–an organism that grows best at high temperatures

thermotolerant–tolerating high temperatures but growing best at lower temperatures

thiobacilli–a group of bacteria that use sulphur as an energy source

tidal heating–heating produced as a result of variation in the gravitational forces acting on a planet or moon

torpor–a period of dormancy

transpiration–the loss of water by a terrestrial plant, usually via the leaves

tundra–treeless regions of the Arctic or mountains where the subsoil is permanently frozen

ultraviolet (uv) radiation–that which has a wavelength shorter than that of the violet end of the visible spectrum but longer than that of X-rays

unfreezable water–see osmotically inactive water

unicellular–consisting of a single cell

unsaturated fatty acids–fatty acids containing one or more double bonds

vertebrate–an animal with a backbone

vitrification–conversion into a glass or glass-like substance

X-rays–electromagnetic radiation of high energy and very short wavelength (between that of ultraviolet and gamma rays)

yeast–a unicellular fungus

Index

Note: page numbers in bold denote illustrations; those in italics, tables.